The Peripatetic Astronomer

Charles Piazzi Smyth. Portrait by John Faed, RSA (Royal Society of Edinburgh).

The Peripatetic Astronomer

The Life of Charles Piazzi Smyth

H A Brück and M T Brück

Adam Hilger, Bristol and Philadelphia

British Library Cataloguing in Publication Data

Brück, Hermann A
The peripatetic astronomer: the life of
Charles Piazzi Smyth.
1. Piazzi Smyth, Charles 2. Astronomers
——Great Britain——Biography
I. Title II. Brück, M. T.
520′.92′4 QB36.P4/

ISBN 0-85274-420-X

Library of Congress Cataloging-in-Publication Data

Brück, H. A. (Hermann Alexander), 1905–
The peripatetic astronomer

Bibliography: p.
Includes indexes.
1. Smyth, C. Piazzi (Charles Piazzi), 1819–1900.
2. Astronomy—Great Britain—History—19th century.
3. Astronomers—Great Britain—Biography. I. Brück,
M. T. (Mary T.) II. Title
QB36.S645B78 1988 520′.92′4 [B] 87-28878

ISBN 0-85274-420-X

Consultant Editor: **Professor A J Meadows**,
Loughborough University

Published under the Adam Hilger imprint by IOP Publishing Ltd
Techno House, Redcliffe Way, Bristol BS1 6NX, England
242 Cherry Street, Philadelphia, PA 19106, USA

Typeset by Keytec, Bridport, Dorset

Printed in Great Britain by J W Arrowsmith Ltd, Bristol

To Sir John and Lady Clerk of Penicuik

Contents

Preface and Acknowledgments

Charles Piazzi Smyth, astronomer, geodesist, spectroscopist, meteorologist, metrologist, pioneer photographer, artist, traveller, writer and pyramidologist was one of the colourful personalities in the world of British science in the nineteenth century. Not all his ventures were fruitful and not all of them can be compressed into the pages of one book. However, he made one supremely important contribution to astronomy which alone entitles him to a place in the history of science, that is, his advocacy of 'mountain astronomy' and his experiment in 1856 on the Peak of Tenerife which demonstrated that there, in Isaac Newton's 'serene air above the grosser clouds' lay the future for observational astronomy.

Piazzi Smyth pointed out the folly of building astronomical observatories in large cities and pleaded for 'peripatetic' astronomers who would place their instruments in suitably selected sites and travel as needed between their home base and their high-altitude station. Today the peripatetic astronomer has become the norm. Observatories at ground level are no longer used for their original purpose but as centres from which astronomers travel to locations in remote parts of the world. In his own lifetime Piazzi Smyth was disappointed in not being able to realise his ambition of making his observatory in Edinburgh the world's first peripatetic institution. He would have found it sad and ironical that a century after his experiment on Tenerife Britain's 98-inch telescope, dedicated to Isaac Newton, should be erected in the low-lying fields of England. When at last in the 1970s it was decided that Britain should have its own mountain observatory the Edinburgh astronomers who were made officially responsible for a search for suitable sites found that Piazzi Smyth's Canary Islands were still among the best in the world, and that Tenerife's unspoilt sister island of La Palma offered an ideal location for the new observatory.

But while he was unsuccessful in getting support for a permanent mountain observatory, Piazzi Smyth practised his own teaching and became a peripatetic scientist, travelling at his own expense and with his wife as sole assistant to make his observations at the best sites he could afford to visit, at Palermo, Lisbon and Madeira.

Piazzi Smyth was brought up in the great age of the natural philosophers, of the experimenters and observers, represented in Britain by people like John Herschel, David Brewster and James Forbes, all of whom were known personally to him. In those days men like Piazzi Smyth who had no formal university education, and amateur astronomers and observers of nature, could play a full part in the progress of science. In the middle years of the nineteenth century these men were ousted by the new young generation, led in Britain by William Thomson, James Joule and James Clerk Maxwell, who combined theoretical with experimental physics and made leaps in the understanding of nature which the amateurs could not hope to grasp. Piazzi Smyth is one of those who fell between these two generations of scientists, and his own story is an example of how one man was able to cope with the changing situation. His solution was to choose a new field of research, spectroscopy, which was then still beyond the reach of theoretical physics; and in the second half of his scientific career he became a leading spectroscopist, both in the laboratory and in observations of the Sun and the Earth's atmosphere.

In his middle years Piazzi Smyth got involved in the study of the Great Pyramid of Gizeh. What began as a metrological exercise took on for him an extraordinary mystical aspect, and he found himself the leader of a movement which spread through many countries and attracted thousands of devotees. Pyramidology is a cult which is still very much alive though not exactly in the form in which Piazzi Smyth expounded it.

There is abundant documentation of Piazzi Smyth's scientific life which spans more than 60 years. His early days as a young astronomer in South Africa have been well recorded in one of Brian Warner's volumes on the history of astronomy at the Cape. Records of his following 50 years of activity are to be found in the archives of the Royal Society of Edinburgh and the Royal Observatory Edinburgh. He bequeathed to the Royal Society of Edinburgh his journals, which span 30 years, several albums of paintings and hundreds of photographs. His numerous laboratory working notebooks, dealing with optics, spectroscopy, and other experimental topics which were also left to the Royal Society of Edinburgh are now preserved in the archives of the Royal Observatory Edinburgh, which also house the official observatory log books for his 42 years as Astronomer Royal for Scotland and a considerable volume of correspondence. While in office, Piazzi Smyth wrote detailed annual reports for his Board of

Visitors which were printed in the *Edinburgh Astronomical Observations*, the official publications of the Edinburgh Observatory in his day. He published more than a hundred papers in scientific journals and he was the author of some lively accounts of his various travels. He himself put his writings on the metrology of the Great Pyramid into the scientific category; these have also been published in the form of books and papers. A large amount of correspondence on the subject of the pyramid is also preserved in the archives of the Royal Observatory Edinburgh.

The main sources of material for this book are Piazzi Smyth's own publications and books, which are listed at the end of the book, and the archives of the Royal Society of Edinburgh and the Royal Observatory Edinburgh. References to these major sources are not individually acknowledged in the text, and unattributed quotations may be assumed to come from one of them. We have also obtained material from the archives of the Royal Astronomical Society, the Royal Society, the Royal Greenwich Observatory, the University of Edinburgh, the Lick Observatory (Mary Lea Shane Archives), and the Clerk Family of Penicuik. The Bibliography gives a list of books which give the historical and general background to Piazzi Smyth's scientific work. We have not quoted specific references in the text to information which can be found in this list. Similarly, when we refer to scientific material in well-known journals we have not considered it necessary to quote the details in full. In this way we have kept to a minimum the notes and references which are given chapter by chapter in the Notes. Most of the illustrations in the book are taken from Piazzi Smyth's own original photographs, drawings or diagrams; reproductions are by the University of Edinburgh Audio-Visual Services, Royal Observatory Edinburgh Photolabs and Murray Donald of Edinburgh.

We wish to acknowledge first and foremost our gratitude to the Royal Society of Edinburgh for generously permitting us to use the Piazzi Smyth archives, and to copy material, photographs and drawings; to the Astronomer Royal for Scotland and the Science and Engineering Research Council for the use of the archives and library of the Royal Observatory, Edinburgh; to the other sources mentioned above, and to the Trustees of the National Museums of Scotland, the Trustees of the Science Museum (London) and the Scottish Tourist Board for supplying photographs. Among individuals who have been particularly helpful it is a pleasure to record our thanks to Professor M S Longair, Mr W H Rutherford, Dr W Duncan, Sir John and Lady Clerk of Penicuik, Dr A Simpson, Professor D E Osterbrock, Professor D G Kendall, Dr J Rösch, Dr P Herring, Dr M Gadsden, Mr A

Macdonald, Mrs A Fraser, Mrs M Fretwell, Mr P Hingley, Mrs Pamela Towlson, Miss Janet Dudley, Mr N H Robinson and Mrs D Schaumberg.

Penicuik, Easter Sunday 1987

Abbreviations used in Notes and captions on illustrations

CPS, Charles Piazzi Smyth
EAO, Edinburgh Astronomical Observations
MNRAS, Monthly Notices of the Royal Astronomical Society
RMS, Royal Museum of Scotland
OI, Piazzi Smyth's *Our Inheritance in the Great Pyramid*
RAS, Royal Astronomical Society
ROE, Royal Observatory Edinburgh
RS, Royal Society
RSE, Royal Society of Edinburgh
RSSA, Royal Scottish Society of Arts

Notes on units

Units have not been standardised. Imperial (feet, inches) and metric (centimetres, angstoms) units of length, all in common use in the nineteenth century, have been retained. (1 foot = 12 inches; 1 inch = 2.54 cm.)

Chapter 1

Piazzi's Godson

Piazzi Smyth was born in Naples on 3 January 1819, into a remarkable family. His flamboyant father, William Henry Smyth, an officer in the Royal Navy, had served with distinction in various actions in the Napoleonic Wars first in Spain and afterwards in the Mediterranean, where he had a command in the Anglo-Sicilian fleet at Messina. After the conclusion of the war in 1815 he had occupied himself for ten years in carrying out surveys of the Mediterranean coasts, which were widely praised, and had published, with support from the Admiralty, a memoir on *The Resources, Inhabitants and Hydrography of Sicily and its Islands*, described as 'the greatest scientific survey ever planned and completed by one individual'. During his time in the Mediterranean Smyth had developed a keen interest in astronomy which was fostered by his friendship with the distinguished Italian astronomer Guiseppe Piazzi, Director of the Palermo Observatory, who had become famous as the discoverer, on 1 January 1801, of the first minor planet Ceres named by him after the ancient goddess of Sicily. Smyth's interest in astronomy received, as he said himself, 'its sharpest spur' when he had the opportunity of assisting Piazzi in the proof-reading of his great *Palermo Star Catalogue*. Astronomy was not the only interest which Smyth acquired in that part of the world; he also cultivated a taste for archaeology and was responsible for collecting in Tripoli some interesting architectural specimens which were presented later to the British Museum. The versatile Smyth was also an accomplished linguist who wrote and published occasionally in French.

While in Sicily Smyth had married Annarella Warington, the only daughter of the English Consul to the Bourbon Kingdom of the Two Sicilies. A woman of exceptional intelligence and charm, she exerted a strong influence on their family of three sons and six daughters, most of whom were to reach high positions in society. Charles, the second

William Henry Smyth (RAS Archives PP9).

child and second son, was christened in the Anglican Church in Naples and was given as his second name that of his father's friend Piazzi who, though a Catholic priest—a member of the Order of the Theatines—gladly acted as godfather to the child. It is said that at the christening Piazzi made a wish that the young Charles Piazzi should become an astronomer like himself. With the support of his enthusiastic father this wish was indeed to be fulfilled in full measure. For his part, Charles throughout his life was proud to add his godfather's name to that of his father, and in his adult life chose to be known as Piazzi among family and friends.

On completing his Mediterranean surveys in 1824 W H Smyth retired from the Royal Navy in order to find more time for the pursuit of his scientific and literary interests. He joined the Royal Astronomical Society soon after its foundation in 1820; he also became a Fellow of the Society of Antiquaries and in 1826 was elected a Fellow of the Royal Society. Astronomy now became the central interest of his life.

At the Royal Astronomical Society he made a circle of new friends of whom the most illustrious by far was John (later Sir John) Herschel. Though Herschel was still only in his early thirties when Smyth first got to know him he was already looked upon with great respect as a senior figure in the world of science. Being a man of independent means, he could represent in the Royal Astronomical Society a focus for both professional astronomers and for the many amateur astronomers who looked for a purpose in their chosen hobby. Soon after the Society's foundation a small group of members formed themselves into a club and regularly dined together after meetings. Herschel was among the first members of this club of which Smyth, always a companionable man, became one of its greatest devotees. In fact he hardly missed a club dinner for thirty years and was its President for a long period. The Royal Astronomical Society and the club coterie placed W H Smyth in a position of considerable influence in the British astronomical community.

Back in England Smyth—later raised to the rank of admiral—moved with his family to Bedford, a town about fifty miles north of London, where he erected an astronomical observatory intended for serious observational work. He equipped this with a number of good instruments including an equatorially mounted six-inch refractor, one of the first in the country to be driven by clockwork and looked upon as a model of what a perfect telescope ought to be. In his 'Temple of Urania', as he called his observatory, Smyth settled down to a programme of observation of double stars, drawn mainly from Piazzi's *Catalogue*, which was to occupy him for nine years. The fruit of this undertaking was the *Bedford Catalogue* of 850 double stars, star clusters and nebulae, which he published in 1844 as part of his *Cycle of Celestial Objects*, with a general introduction to the history and practice of astronomy which created an enormous popular interest in the subject. It was the first book of its genre, an enjoyable and useful text for non-professionals in that 'golden age of amateur astronomers'.[1] In 1845 the book gained for its author the Gold Medal of the Royal Astronomical Society; a year later Smyth reached the pinnacle of his ambition on being elected President of the Royal Astronomical Society.

Meantime young Piazzi and his brother Warington attended the excellent local school, Bedford Grammar School. However, unlike his brother who went to Cambridge and eventually became a distinguished geologist, Piazzi did not proceed to university. He left school at only sixteen years of age on being made an offer of an astronomical post and the prospect of a career in his favourite subject. By this stage in his education he had received a thorough grounding in the classics, while at home he had acquired artistic skills from his gifted mother and had also learned modern languages; in later life he could freely read French, Italian and German. In mathematics he had been taught his Euclid, but by cutting short his schooling he was left at a considerable disadvantage as regards mathematical training, a lack which in spite of his best endeavours he was never able to remedy.

The offer came from his father's friend Thomas Maclear, His Majesty's Astronomer at the Royal Observatory, Cape of Good Hope, a post to which he was appointed in 1834 and which he was to occupy for 46 years. The work of the Cape Observatory had not been very successful in its early stages, and when Maclear took over he insisted on the early appointment of a capable assistant. Knowing the Smyth family well he calculated that their schoolboy son could under his guidance become a competent 'coadjutor' in spite of, even because of, his youth. He proposed that Piazzi should become his assistant at the Royal Observatory at the Cape of Good Hope with a generous salary of £250, which was only £50 less than he was to get years later in his professorial post in Edinburgh!

Young Piazzi Smyth, whose childhood and boyhood had been spent among his father's astronomical instruments, was indeed a good choice for the task which Maclear had in mind. At the Cape Observatory the principal programme of work was concerned with the determination of positions on the sky of the brighter stars which were visible from the southern hemisphere. The work involved the precise timing of the passages of stars through the meridian and the determination of their altitudes above the horizon at the time of their meridian passage. The actual observations were carried out according to standard practice with precision clocks and telescopes which could be moved in the meridian only. 'Grinding the meridian', as the work was sometimes called, demanded little intellectual prowess, but depended on a certain degree of alertness and skill, and on some familiarity with the imperfections of scientific instruments. Likewise the necessary computational techniques could be learnt without any academic knowledge of astronomy or mathematics. Maclear's offer was accepted, and Piazzi Smyth left England in the summer of 1835 and after a voyage of 59 days arrived in Table Bay in October.

He was warmly welcomed at the Cape by the Maclears and their large family of children, the eldest only eight years of age, who were to remain good friends all his life. However, kind though Maclear was in his domestic circle, he turned out to be a very stern director in the observatory; he confessed to making his one and only assistant work for 18 hours of the day. Piazzi Smyth did not complain but performed his astronomical duties with great enthusiasm and according to Maclear himself was 'as regular at his post as the Transit Clock'. He made the required observations of meridian passages of stars as carefully as he carried out the tedious calculations for converting the raw observational data into actual star positions. Piazzi Smyth became responsible for the operation of a time-ball which Maclear installed in 1836 for the benefit of shipping in Table Bay; he was also given the tiresome task of checking the proof sheets of the Cape Observatory's first *Star Catalogue*. There was little or no time left for reading or for study.

Entirely new responsibilities fell on the Cape astronomers when the observatory was asked by the Admiralty to embark on a major geodetic programme and re-measure the exact length of an arc of the meridian running through Cape Town. The survey in which Piazzi Smyth took a very active part was to take a total of twelve years from 1836 to 1848. The work was arduous once operations moved away from the neighbourhood of Cape Town and up to the mountainous regions to the east and north of the observatory, which were at all times difficult to reach and particularly so when they were snow-capped in winter. Piazzi Smyth's robust physique, which he had inherited from his father, helped him to emerge triumphant from all his difficulties and to sustain the morale of his helpers in the surveying team. He was involved in the survey at intervals over a period of five years, including the entire winter of 1844–5, his last in South Africa, which he spent on Sneeukop in fearsomely harsh conditions of wind, mist and ice. While working in the mountainous regions of the Western Cape away from all human habitation he was greatly struck by the brilliance of the stars in the sky and by the visibility of faint objects such as the glow of the Zodiacal Light, a phenomenon in which he remained strongly interested all his life.

During Piazzi Smyth's early years in South Africa John Herschel was also there, working at the temporary observatory which he had set up with his 18-inch 20-foot 'front view' reflecting telescope at Feldhausen, six miles from Cape Town. His purpose was to supplement with observations in the southern hemisphere the survey of the northern sky which his father, Sir William Herschel, had carried out from England. Piazzi Smyth had only just arrived at the Cape Observatory when Comet Halley became visible in South Africa. He succeeded in producing a set of sketches of the comet which according to Maclear gave a quite outstanding account of the changes in its appearance. These observations were of particular interest to John Herschel, who observed the comet after its perihelion passage in 1836 with his great reflector in Feldhausen and who was then beginning to think deeply about the causes for the varying structure of the tails of comets. The great daylight comet of 1843 had an unusually long forked tail of which Piazzi Smyth produced for the benefit of Herschel 'a fine drawing' which received special mention in Herschel's *Outlines of Astronomy*. That comet tempted Piazzi Smyth to stray further beyond the 'grinding of the meridian'. Using an old reflecting telescope belonging to the observatory and a piece of rock crystal in the light beam Piazzi Smyth tried to find out how much of the comet's light might be due to simple scattering of sunlight. His search for the presence of polarisation in the comet's light led to no definite result, but it hardly deserved the strong disapproval of Maclear who made it very plain that for him 'proper astronomy' was concerned with positions and motions of celestial bodies and definitely not with what he termed 'physical descriptions'.

While in South Africa, Herschel had been much impressed by Piazzi Smyth's activities both as an astronomer in the observatory and as a geodesist in Maclear's trigonometrical survey. He recommended to the Smyth parents that they should encourage their son to improve his mathematical knowledge, and on Sir John's advice some 'hard' books were sent out to Piazzi to study in his leisure time—'if', as his mother remarked, 'he can be said to have any'.[2] For his part, Piazzi, in those early years at the Cape, developed a hero-worship for Herschel which persisted throughout his entire life.

Another link between Herschel and Piazzi Smyth followed the discovery in the summer of 1839, mentioned by Herschel in a letter to Maclear, of the photographic process of J L M Daguerre. Herschel was intensely interested in the possibilities opened up by the new process and soon succeeded in his own attempts to fix photographic images on suitably sensitised paper. He worked along the lines which Fox Talbot had described in 1834 in his work on 'photogenic drawings' which he patented in 1840 as the process of 'Talbotype' or 'calotype'. The great photographic discoveries created a major new interest for Piazzi Smyth, who by the time that Herschel was able to publish technical details of his work in 1842 had already started to experiment on his own. As early as 1843 Piazzi Smyth succeeded in producing what are the oldest known photographs, all calotypes, of people and scenes in South Africa. His pictures include one of the Cape Observatory which may well be the earliest photograph of any observatory in the world.

Sir John Herschel in 1836 (RAS Archives PP4B).

Piazzi Smyth was away from Cape Town engaged on survey work when notice was received in England of the untimely death in his 46th year of Thomas Henderson, first Astronomer Royal for Scotland. When Piazzi Smyth's parents, always ambitious and devoted on behalf of their children, heard the news, they considered that their son was sufficiently well trained in Henderson's field of positional astronomy to be capable of succeeding him. Knowing that the news of the Edinburgh vacancy was not likely to reach South Africa in time to allow their son to submit an early application, Smyth took it upon himself to apply for the vacant post in his son's name. The parental move was not as presumptuous as it might seem. In the years since Piazzi had been away his father had reached a position of considerable eminence in British astronomy. As President of the Royal Astronomical Society and member of the Board of Visitors of the Royal Observatory, Greenwich, he felt justified in alerting those friends who were familiar with his son's 'energetic industry' in South Africa and in asking them to support his application. The most influential among his friends was, of course, Sir John Herschel who in a letter to Smyth on 31 December 1844 promised his support for Piazzi's candidature.

The importance and the particular nature of the joint government–university post in Edinburgh delayed a decision by the selectors for three months, but in April 1845 W H Smyth was informed by the Home Secretary, Sir James Graham, of his son's appointment. Oddly enough, Maclear had not been informed of the father's vicarious application. He received the news in June from Sir Francis Beaufort by the same mail by which Piazzi Smyth himself learned of it. Though surprised, he was generous enough to understand the particular circumstances in which the appointment had been made and accepted the supporting role played by Sir John Herschel in the matter.

Piazzi Smyth himself received the news on 10 June on his return to the observatory from his field work, but he conscientiously insisted on completing his part in the survey before considering his new duties. In fact, he returned to the survey only three days after learning of his appointment, and it was only in August, two months later, that Maclear could prevail on him to hand over his survey duties to his colleague William Mann and to return to Cape Town to prepare for his departure to Edinburgh. Having said farewell to his many friends in South Africa, Piazzi Smyth embarked at last in the second half of October on the *City of Poonah* for Gravesend, where he arrived in the middle of December 1845.

The ten years which Piazzi Smyth had spent as assistant at the Cape Observatory had been extraordinarily full. He had of course learned to the highest degree of meticulousness the practical skills of meridian astronomy and geodesy and had contributed in the fullest measure to the work of the observatory and the survey. But his activities went far beyond those duties; he recorded for posterity in his drawings the operations of the survey; he was a pioneer in the new science of photography; he was something of an inventor, and an imaginative astronomical observer. This intensely full and fascinating period of his life as a very young 'astronomer–artist' has already been made the subject of a book by Professor Brian Warner.

The Smyth family were no longer living in Bedford when Piazzi returned home. When his father had completed the catalogue for his *Cycle* he donated his beloved equatorial to a neighbour and friend, Dr John Lee, another convert to his astronomical enthusiasm. Lee was a man of independent means who lived on the estate of Hartwell in Buckinghamshire, where he had erected under Smyth's guidance a proper astronomical observatory with Smyth's equatorial as its main instrument. Lee could afford to run his observatory on professional lines and to employ a director to take charge of the observational work. Among Lee's astronomers was Norman Pogson, later Director of Madras Observatory, who will always be remembered as the man who first defined the magnitude scale for recording stellar brightness.

Hartwell House. Drawing by Henry Augustus Smyth (Aedes Harwellianae).

The Smyths, after some years in London, moved to a house called St John's Lodge in Aylesbury, close to Hartwell, where Smyth could indulge his interests and the families could cultivate their friendship. Lee, who was an antiquarian and a collector of ancient coins and works of art, found a fellow enthusiast in the Admiral—as he was by then—who having finished his astronomical catalogue undertook an

John Lee of Hartwell in his Meridian Room by CPS (RSE).

inventory of the Hartwell House Museum. Piazzi Smyth knew Lee from his childhood, and on his return from South Africa found in him a good friend and generous patron.

Piazzi spent a happy month with family and friends in England and attended the meeting of the Royal Astronomical Society and the club dinner on 9 January 1846 as the guest of his proud father. After ten years away from home he had to find his place in the family again. Both of his brothers were at the time in Ireland, the elder, Warington, already an established geologist with the Geological Survey, and the younger, Henry, Piazzi's junior by six years, with the Army. Piazzi looked to Warington for advice: 'You have been so successful in all that you have undertaken and have won such golden opinions from all sorts of persons that your notions on things in general must be well worth learning and studying by one who has all to learn still.' They had been very close as boys and were able to recover their particular affection for each other which remained for the rest of their lives. The eldest sister, Henrietta, was soon to marry the Rev. Baden Powell, Professor of Mathematics at Oxford, a middle-aged widower with four children. The Smyth family were very proud of this connection: writing to his sister Piazzi referred to Baden Powell as 'the light not only of Oxford but of England in general'. Henrietta was an intellectual lady, interested in music and in science, later to make a name for herself as the mother and inspiration of the formidable Baden Powell family.[3] The Smyths were to experience sadness in the midst of the excitement. One of the younger sisters, who was to keep house for Piazzi, fell ill after little more than a month in Edinburgh and returned home with her mother. She died, aged 19, in January 1847.

Piazzi Smyth arrived in Edinburgh at the end of January 1846. He had been ten years away from Britain, having left as a young, shy and eager schoolboy and returned a university professor.

The Royal Warrant signed by Queen Victoria appointed Piazzi Smyth to the post of Regius Professor of Practical Astronomy in the University of Edinburgh and Astronomer Royal for Scotland. The Warrant enjoined on him 'to take upon himself the care and custody of all instruments within the Observatory of Edinburgh which belong to Us, and to apply himself with diligence and zeal to making astronomical observations at the said Observatory for the extension and improvement of Astronomy, Geography and Navigation, and other branches of Science connected therewith'. The text referred to the agreement whereby the 'Members of the Astronomical Institution of Edinburgh had given the use of the Observatory at Edinburgh erected by them, and the instruments therein contained, to the Professor of Astronomy in the University.'

The style of the Royal Warrants for the appointment of Edinburgh's Astronomers Royal has remained practically the same until the present day, but the particular wording in the case of Piazzi Smyth's appointment was rather unusual. It drew together three different threads: the University of Edinburgh which administered the Chair of Practical Astronomy; the Astronomical Institution of Edinburgh, the founders and legal owners of the observatory building; and the Government, who owned the instruments, paid the salaries and provided general maintenance.

The University's Regius Chair of Practical Astronomy had been originally founded in 1785 to fill a need in the teaching of nautical astronomy in a city which adjoined the thriving port of Leith. In practice, since the first holder of the Chair had refused to teach, the post turned out to be a sinecure. The Astronomical Institution of Edinburgh, founded in 1811 as an offshoot of the Scottish Enlightenment, was a remarkable society created by private citizens with the support of town and university. Preceding the foundation of the Royal Astronomical Society in London by nine years, the Edinburgh Institution was in fact the oldest astronomical society in the United Kingdom. The Institution counted among its members many of the leading citizens of the day; its President was Professor John Playfair who at the time of the foundation held the Chair of Natural Philosophy in the university; and its Vice-President was Sir George Steuart Mackenzie, the noted mineralogist and traveller. A major aim of the Institution was the establishment of a proper observatory which could be used both for scientific research and for popular instruction. Funds were raised by private subscription for the creation of a 'scientific observatory' where 'accurate observations could be made with the best procurable instruments' and also of a 'popular observatory and physical cabinet' where members of the Institution could make 'observations and experiments in astronomy or any other science'. As a site for the proposed two observatories the Calton Hill near the centre of the city was chosen—the place where Thomas Short, a brother of the telescope maker James Short, had earlier endeavoured to establish an observatory. The members of the Astronomical Institution agreed to use the remnants of Thomas Short's 'Old Observatory' for their popular observatory and to build their new scientific observatory beside it on the same site.

A plan for the building of what came to be known as the 'New Observatory' was entrusted to Playfair's nephew, the young architect William H Playfair, who was to become responsible for the design of many of Edinburgh's fine classical buildings. Playfair designed the observatory as a cruciform Roman Doric structure with six pillars in front of each of its four strictly equal sides, and a prominent dome for a telescope in its centre. His unusual and striking plan was readily accepted by the members of the Astronomical Institution and the foundation stone of the new building was laid on 25 April 1818 by Sir George Mackenzie. The lofty hopes of those present on that occasion were engraved on a platinum plate, '*ne diutius urbi clarissimae scientiam omnium pulcherrimam atque amplissimam excolendi facultas deesset*' (that a renowned city should no longer lack the facilities for the pursuit of the fairest and grandest of the sciences). At the corner of the outer wall was placed a monument to the memory of his uncle by the architect. Playfair's charming building survives unchanged to this day.

The substantial building costs of the new observatory entirely exhausted the funds of the Astronomical Institution so that when it came to the acquisition of instruments the members had to appeal to the Government for support. Their request for assistance was answered by a special grant of £2000. The interest of astronomers at the time was centred on the determination of the positions of stars on the celestial sphere. The determination of their two spherical coordinates, right ascension and declination—somewhat comparable to longitude and latitude on the terrestrial sphere—was carried out in practice with telescopes mounted in the meridian: transit instruments and clocks for the timing of meridian passages of stars and the measurement of right ascension, and mural circles for the measurement of declination. Two such telescopes were accordingly acquired by the Astronomical Institution out of their government grant. The first was an eight-foot transit instrument made by the celebrated firm of Repsold in Hamburg; the second was a six-foot mural circle made by the London firm of Troughton and Simms, an instrument very similar to the one at the Royal Observatory at Greenwich. The two meridian telescopes continued to constitute the main equipment of the observatory for half a century. There were some smaller instruments donated by members of the Institution and an exceptionally fine altitude and azimuth instrument made by Troughton and Simms which had been presented to Playfair by students in his astronomy class and bequeathed to the Royal Observatory on his death in 1819. All these instruments were placed under the supervision of William Wallace, Professor of Mathematics in the university who acted as honorary 'Observer' for the Astronomical Institution. In August 1822 when King George IV paid his celebrated visit to Edinburgh, the Astronomical Institution having presented a loyal address to the King received the royal accolade and their observatory was honoured with the title of 'Royal Observatory of King George IV'.

On the death in 1828 of Robert Blair, the inactive first holder of the Chair of Practical Astronomy, the university authorities found themselves in a quandary. A Royal Commission on the affairs of the Scottish universities was sitting at the time and when they came to a discussion of the Chair of Practical Astronomy they recommended that it should not be filled 'until a suitable observatory attached to the University could be provided'. Discussions were started about the future position of astronomy in Edinburgh between the Astronomical Institution, the University and the Government which led in 1834 to a formal agreement whereby the Astronomical Institution would make over to the University 'for their unlimited use' the Institution's 'Scientific Observatory' on condition that the Government would step in and convert this observatory into a properly supported public establishment with a Principal Observer at its head and a second astronomer to assist him. It was further agreed that the office of Principal Observer should be combined with that of the Regius Professor of Practical Astronomy and that the holder of the joint post should in future have the title 'Astronomer Royal for Scotland and Professor of Practical Astronomy in the University of Edinburgh'. When the University investigated the potential demand for astronomical tuition they concluded that student numbers were unlikely to exceed six in any one year, and they therefore also decided that the university post should remain honorary with the professor devoting himself mainly to his observational duties.

The first holder of the combined post, Thomas Henderson, was a most suitable choice. In 1832 Henderson had become Director and HM Astronomer at the Royal Observatory at the Cape, but had resigned the post and returned to his native Scotland only one year later. Henderson received his appointment to Edinburgh in 1834 when he became, with a Royal Warrant signed by King William IV, the first Astronomer Royal for Scotland and the second Professor of Practical Astronomy in the University of Edinburgh. Following his appointment Henderson lost no time in starting a major observing programme concerned with the determination of the positions of the Sun, Moon and planets, and of as many of the brighter stars as possible. All the observations with the mural circle were made by Henderson himself, those with the transit circle by his assistant, Alexander Wallace, a graduate of Edinburgh University. Incredible as it may seem, Henderson and Wallace managed to collect in ten years no fewer than 60 000 observations of star positions, to make the necessary computations for about half of them and to publish the results in the first five substantial quarto volumes of the series of the *Edinburgh Astronomical Observations*.

The huge amount of work involved in the Edinburgh programme did not hinder Henderson from carrying on with the reduction of the observations which he had made at the Cape. The most important of these was the discovery of stellar parallax and the measurement of the distance of the star Alpha Centauri, the third brightest star in the sky and a prominent object at the Cape. It was unfortunate for Henderson that he delayed the publication of his result and was anticipated—by just three months—by F W Bessel, the Director of the Observatory at Königsberg in Eastern Prussia, who succeeded in measuring the distance of the star 61 Cygni. The precedence of Bessel's public announcement over Henderson's did not diminish the credit which Henderson has been rightly given in the history of astronomy as the co-discoverer of stellar distance.

Henderson died in November 1844 at the early age of 45 after only ten years in his Edinburgh post. This was the distinguished and dedicated man whom the 27-year-old Piazzi Smyth came to replace.

Chapter 2

Astronomer Royal for Scotland

The Calton Hill, the location of the Royal Observatory which became Piazzi Smyth's charge in 1846, is a conspicuous landmark near the centre of Scotland's capital. Few would disagree with Robert Louis Stevenson when, in his *Picturesque Notes*, he described the view from the Calton Hill as the finest in the city. In spite of seeing it for the first time in the depth of winter, Piazzi Smyth was overwhelmed by the beauty of the panorama which he described in his own effusive style to his family and friends. 'My own adopted town', he called Edinburgh in a letter to Lady Herschel written a few weeks after his arrival, 'a city on the one hand so severely classical and on the other so boldly, irregularly picturesque, stretching its white arms to the sea. Had there been no town built on the site of Edinburgh it would still have been a strikingly romantic spot; with its craggy precipitous hills and ravines and chasms; and distant views beyond of sea and fell. But when the same eminences are crowned with magnificent buildings, here the severe simplicity of classic taste, there with all the complexity of barbaric ornament of the Middle Ages, a spectacle is beheld such as perhaps no other city in the world could offer.' Many of Piazzi Smyth's sketchbooks survive from those early days, depicting scenes of Edinburgh and of the Scottish countryside which bear witness to the love for his adopted country which stayed with him for the 40 years he was to live there.

Piazzi Smyth was equally enthralled by the intellectual climate of the northern city which in the eighteenth century had become one of the world's great centres of learning. The city with its university and its learned societies was a revelation to a studious young man whose experience of academic life had up till then been very limited; and the status accorded to his prestigious post must have seemed to him a striking contrast to the subservient role which he had played as

Calton Hill Observatory. On the left is the Old Observatory and on the right the memorial to John Playfair (ROE).

'coadjutor' at the Cape. In this exciting milieu Piazzi Smyth was received with great warmth by his illustrious new friends. Waiting to greet him on his arrival and to welcome him into his family circle was Professor James D Forbes who had looked after the affairs of the Royal Observatory in the year which had elapsed since Henderson's death. Forbes was an outstanding physicist, regarded by his contemporaries as one of the luminaries of the age, and a close friend of George Airy, the Astronomer Royal. Forbes' closest colleague in the university, the Rev. Philip Kelland, the genial Professor of Mathematics, was also among the first to welcome Piazzi Smyth in his home.

Forbes introduced the young newcomer to other members of the university and to the principal figures in the affairs of the observatory. Within a month of his arrival Piazzi Smyth was inducted into the Senatus Academicus which had among its members many distinguished literary men such as John Wilson ('Christopher North', the writer of *Noctes Ambrosianae*), Cosmo Innes, lawyer and historian, and E Aytoun, Professor of Rhetoric and English Literature. There was also Thomas Thomson, a member of the Cosmo Innes circle and a keen member of the Astronomical Institution. These and other names were mentioned with admiration in letters which Piazzi Smyth wrote to his family and friends. From the first he cultivated the arts and classical scholars quite as much as he did the scientists. He was particularly happy with the well-stocked university library: as he wrote to his old friend William Mann at the Cape, 'I am fortunately situated having almost unlimited fingers in the College Library.'

Piazzi Smyth had scarcely arrived in Edinburgh when he was proposed by Sir Thomas Makdougal Brisbane, the President, as a Fellow of the Royal Society of Edinburgh and, after only six weeks in his post, duly elected. The previous week he had travelled to London to be admitted as a Fellow of the Royal Astronomical Society by his father who was the Society's President, a proud occasion for father and son and all the family: Piazzi and his sisters used to refer grandly to their father as 'The President'. In his elevated rank the youthful Astronomer Royal also found himself enjoying an active social life in Edinburgh where he was placed on the High Commissioner's guest list for a soirée at Holyrood Palace. He made friends with local families, one of the first being that of the Clerks of Penicuik who at their family seat some ten miles outside Edinburgh had been leaders in the scientific and cultural life of Scotland for many generations. Henry Clerk, son of Sir George Clerk, the senior member of the family, and a cousin of James Clerk Maxwell, had been attached to the Magnetic Station at the Cape where he became a good friend of Piazzi's and of his colleague Mann. Clerk had been responsible for the magnetic observations during the survey of the Antarctic on board the

Pandora which sailed in 1845 from South Africa to Australia along the southern 60th parallel of latitude. He had returned to Britain at the same time as Piazzi Smyth, and was stationed with the Army at Woolwich where he later became a colonel and finally a general. Henry Clerk remained a close friend of Piazzi Smyth who often enjoyed the hospitality of the Clerk family in their beautiful Penicuik House.

Piazzi Smyth also found himself the occupier of an impressive official residence which had been specially provided for his predecessor Henderson. The house, 1 Hillside Crescent, which still stands, was at the end of an elegant terrace designed by William Playfair in Edinburgh's New Town and was situated at the foot of the Calton Hill, a short walk from the observatory. Piazzi Smyth described it to the Herschels as large, comfortable and magnificent, and, with the assistance of his mother, he furnished it in a manner appropriate to the mode of life of a serious-minded, God-fearing scientist.

Enthusiastic though Piazzi Smyth was about his new surroundings and impressive though the title of his office sounded, the reality of the situation he met with when he took over the work of the Royal Observatory was daunting: a building in a sad state of neglect and a pile of some 30 000 unreduced observations accumulated by his predecessor during the previous six years. So numerous were the observations that, as Piazzi Smyth remarked, 'one would think that they had been made in some sunny southern climate, so few are the days when something was not done with the instruments'. Henderson had been a single-minded astronomer of the old school who saw it as his only duty in life to record nightly every star and planet that crossed the meridian. This task he had performed with the help of his one selfless assistant, Alexander Wallace, whose association with the observatory went back to 1834, two years before the appointment of Henderson. Piazzi Smyth on his arrival found him hard at work on the reductions.

Henderson's name was revered throughout the astronomical world, coupled as it was with that of Bessel in the historic discovery of stellar distance, and it was natural that Piazzi Smyth should see it as his first duty to preserve for posterity the positional observations which Henderson had left unreduced. His first decision on taking up office, therefore, was to reduce and publish the remaining Henderson and Wallace observations before embarking on any fresh programme of work, even if this meant suppressing his own ambitions and perhaps giving the impression to the outside world of not doing any work himself. He outlined his plan to Herschel, saying that it seemed the only chance of Henderson's work 'seeing the light' and that in any case 'it would be the more economical procedure to complete observations already made than to make more'.

Though Piazzi Smyth foresaw that the completion of the old reductions might well occupy him for five years, he did not delay in considering future astronomical programmes which might be carried out afterwards. The observatory, he argued, should not attempt to compete with major centres of meridian astronomy like Greenwich or Paris; but he contended that a publicly supported institution ought to have a long-term programme with a specific goal of a kind which would be beyond the reach of private observatories. He decided that such a programme might be a systematic study of stellar proper motions with the aim of securing a reliable determination of the motion of the Sun through space. That the Sun moved through space like other stars had been discovered by William Herschel as far back as 1783 from the effect of its motion on that of only seven other stars; F W A Argelander of Bonn had in 1837 amplified the observations to include as many as 390 stars. Though this was a major advance over William Herschel's work, Piazzi Smyth pointed to the limitations of Argelander's data which had been derived from observations of star positions given in catalogues published by different observatories and based on data spread unequally in time and reduced on different assumptions. Piazzi Smyth rightly suggested that 'it was very difficult to piece such bits properly together'. He also criticised the fact that only bright stars had been looked at so far, implying that brightness was an indication of proximity, when in fact Argelander had already discovered that the star with the fastest known motion was far from bright having a magnitude of only 6.5. Piazzi Smyth was certain that if the question of the Sun's motion in space was to be settled properly, the motions of many faint as well as bright stars over definite periods of time ought to be scrutinised. 'The actual practice might consist in observing 200 stars, ten times in each one year; taking another 200 stars similarly the next year, and so on for, say, ten years. In the next ten years the observations of the first 10 are to be repeated, and again in a third ten. So that after thirty years safe and secure ground for discussing the cynosure of modern astronomy would be secured.' The use of three epochs might reveal whether the motions were uniform, as they were generally assumed to be, or not. The possibility that the stars move in circular orbits round the centre of gravity of the revolving stellar multitude had in fact been suggested in the very year that Piazzi Smyth discussed his stellar motion programme, by the German astronomer J H Mädler, the successor of Otto Struve as the Director of the observatory of Dorpat (now Tartu, Estonia).

Piazzi Smyth laid his well-thought-out programme before Airy and Herschel who, according to correspondence with the Rev. R Sheepshanks,[1] Secretary of the Royal Astronomical Society, fully approved, though Sheepshanks himself thought the programme 'dull'. Herschel seemed unclear as to whether Piazzi Smyth planned 2000 observations or observations of 2000 stars. 'Would it not be well' he wrote, 'to choose a set of stars which shall contain two or three hundred specially chosen as *not* watched by other observers?'[2] Perhaps Herschel felt that the young astronomer was undertaking too much and was gently trying to suggest a more modest programme. But Piazzi Smyth was determined to observe some 3000 stars over a period of 30 years and in fact doggedly pursued that work from the time the Henderson reductions were completed until his own retirement by which time the positions of no fewer than 3890 stars had been observed, reduced and published. That his programme was very much in keeping with Herschel's own ideas is evident from the latter's discussion on the possible rotation of the Galaxy in the *Outlines of Astronomy* of 1849:

> The question might be at once resolved by the assiduous observation of a considerable number of stars of the Milky Way, judiciously selected for the purpose and including all magnitudes down to the smallest distinctly identifiable and capable of being observed with normal accuracy; we would recommend the enquiry to the special attention of the Directors of permanent observatories; 30 or 40 years of observation perseveringly directed to the object in view could not fail to settle the question.

Herschel and Piazzi Smyth were both over-optimistic in this estimate. The rotation of the Galaxy was not revealed until Jan Oort's analysis of stellar motions in 1927.

The future direction of the observatory's astronomical work was not Piazzi Smyth's only concern when he arrived in Edinburgh. He was worried also by the confused state of its administrative and financial position. The members of the Astronomical Institution were still the legal owners of the building but the Institution was now moribund; many of its founders were dead, others had lost interest or were under the impression that when the Royal Observatory had received a government endowment in 1834 it had passed entirely into government ownership and responsibility. No repairs of any kind had been carried out for 20 years because the whole of the observatory's income was used up on salaries and day-to-day running expenses.

Putting up an optimistic front to Herschel Piazzi Smyth had described the observatory as 'a building which combined a remarkably handsome architectural exterior with a very useful and convenient interior', but he later admitted in a letter to Herschel's daughter Bella that its roof was leaking and that water had to be caught in 'buckets, jugs and basins to prevent a great mess being made on the floor'. It was essential that the observatory building should be legally transferred without delay to government ownership and responsibility. The members of the Astronomical Institution were on the whole willing to renounce their claim and to put the onus of maintenance on the government, but 'delicate manoeuvering', as Piazzi Smyth called it, was needed to placate all concerned. The financial affairs of the observatory, left unclear by Henderson's sudden death, had also to be sorted out.

Two important members of the Astronomical Institution were, however, at hand to help with the legal side of the negotiations, Sir Thomas Makdougal Brisbane and Sir George Mackenzie, both founder members. Brisbane, a famous soldier, had been at one time colonial Governor of Australia where his name is perpetuated in the city of Brisbane. While in Australia he had founded an observatory at Parramatta close to Sydney which supplemented the work of the Cape Observatory. On his retirement to Scotland in 1826 he had set up another observatory at Makerstoun near Kelso in the Scottish Borders which specialised in observations of terrestrial magnetism. Such was Brisbane's reputation that he was elected President of the Royal Society of Edinburgh in 1832 in succession to Sir Walter Scott, a position which he held until his death in 1860 at the age of 87. He was at the same time the President of the Astronomical Institution and as such on close terms with the leading astronomers of the day including Herschel and Piazzi's father. Piazzi's parents visiting Edinburgh shortly after the appointment of their son had a discussion with Brisbane which led to the solution of many of his problems. The unseen hand of his parents and the influence of their friends in high places had helped as they had done before. It was 'Mamma' with her eye on practical matters who prompted Piazzi to press at the same time for a reduction in the feu duty[3] to be paid on his residence; from then on he lived rent-free as part of his perquisites of office.

Meanwhile W H Playfair, the observatory's architect, was called in to assess the structural state of the building. The estimate for restoring the observatory was £500 and there was also need for repairs to the official residence. The Institution, unwilling to undertake these or any future expenses agreed to put a formal request to the government to take over the observatory entirely. They recommended that the observatory should be made over to the government under the conditions

> That the office of Astronomer Royal be ever hereafter associated with the Professorship of Practical Astronomy at the University of Edinburgh,
> That the Astronomer be responsible solely to the Department of Government by which he is appointed,
> That it be recommended that in order to secure the permanent efficiency of the office, a Board of official and other persons connected with Scotland be appointed, including those who have taken the most prominent interest in the proceedings of the Astronomical Institution, to constitute a Board of Visitors, similar to those acting at Greenwich and elsewhere, who shall annually make their report and suggestions to Government on the state of the Observatory.

The document presented to the Lords of the Treasury in which these conditions of transfer were expressed ended: 'The Memorialists beg leave to add, that, if the contemplated arrangement should be effected, they would regard it as the hopeful prelude of a still more liberal patronage of this scientific establishment than it has ever yet received while it remained in their hands.'

All these negotiations were successfully completed by Piazzi Smyth within barely half a year of taking office. He now felt that he could look forward with confidence to securing for Scotland a national observatory which could hold its own among those in other countries of Europe. 'Let the Establishment only once be made a National one', he wrote, 'and it must then, for the honour of the Kingdom (of Scotland), be placed and kept on a foundation suited to the times.' He also approved of the idea of keeping the 'Popular Observatory and Physical Cabinet' in existence for the benefit of members and of the general public. Unfortunately Piazzi Smyth failed to insist on a dependable annual budget which would match these estimable plans.

The new administrative arrangements included the creation of a Board of Visitors on the Greenwich model. A Board superintending all the various government observatories in the Empire had at one time been proposed by the Home Secretary, Sir Robert Peel, who having asked Airy's opinion on the matter had been told that 'only Cambridge men are qualified for being Visitors to an Observatory'. Needless to say such an idea would never have been acceptable to Edinburgh University or to the outgoing owners of the Calton Hill Observatory, let alone Piazzi Smyth. However, Piazzi Smyth was not altogether happy either with the composition of the Board first proposed for him under the new transfer arrangements because it included—in addition to representatives of the university, the

Astronomical Institution and the Royal Society of Edinburgh—three members nominated by the Town Council. While one member, the Lord Provost, would have been acceptable, 'what would be the public opinion of the Greenwich Visitors if the Lord Mayor and the Aldermen had a large voice in its deliberations?', Piazzi Smyth asked; would it not be a case of 'the unlearned elected into judges of science?'. The fact that the Town Council were in dispute with the University at the time was a warning against allowing them to interfere in the observatory's affairs. Supported by Forbes, Piazzi Smyth succeeded in having the town's representation reduced and balanced by including, *ex officio*, the only two people whom he considered competent to be really useful members, the Professors of Natural Philosophy and Mathematics.

The first Board of Visitors, convened in December 1848, reflected the various interests of the cultural centre which Edinburgh was. It consisted of the Principal of Edinburgh University, the President of the Royal Society of Edinburgh, the Lord Provost, the Professors of Natural Philosophy and Mathematics, the Principal Clerk of the Court of Session Thomas Thomson, the Advocate J T Gibson-Craig, Dr David Maclagen MD and the publisher Adam Black. Forbes was elected Chairman of the Board and Kelland Secretary. At their first meeting this Board unanimously recorded their 'strong expression of satisfaction at the zeal, activity and skill displayed by the Astronomer in his arduous duties'.

At the same meeting the members discussed what was to be the first tangible contribution of the observatory to public service, the erection of a time-ball on Calton Hill which would supply time signals to both the city of Edinburgh and the port of Leith. The first ever time-ball had been erected at the Royal Observatory in Greenwich in 1833. The only other one in the United Kingdom served the busy port of Liverpool but Piazzi Smyth was thoroughly familiar with the working of time-balls from his experience at the Cape Observatory, and on his arrival in Edinburgh had promptly investigated the setting up of an efficient time service. Matters had improved but little since the days before the foundation of the Astronomical Institution when 'on the occasion of the arrival in the Firth of Forth of a foreign ship of war, for the express purpose of regulating her chronometers, to the utter amazement of those distinguished strangers, there was not to be found a single man within the city who could tell them what o'clock it was'. Even when Piazzi Smyth arrived in Edinburgh citizens wishing to correct their watches had to ascend Calton Hill to compare their time with that shown on a clock in a window of the observatory. The Tide and Harbour Commissioners by now felt strongly the lack of a service

View of the Calton Hill showing on the right the Nelson Monument with its time-ball and on the left the observatory buildings. (Courtesy of the Scottish Tourist Board.)

for communicating time to the 150 chronometers in ships sailing from the Firth of Forth, and accurate time was also required in the railway stations.

Piazzi Smyth suggested that a perfect location for a time-ball could be found on top of the nearby 100-foot high Nelson Monument, a structure shaped in the form of an upturned telescope which had been erected on Calton Hill in 1807. From this position a ball would be clearly visible from the Firth of Forth and from much of the city of Edinburgh, and the Town Council were prepared to allow the monument to be used for this purpose. The Board of Visitors was in entire agreement with Piazzi Smyth about the scheme and a request for a sum of £200 was sent to the government to meet the cost of the erection of the ball. It took several years, unfortunately, before that money was actually granted. At last, in 1852, a time-ball came into operation which was raised to the top of the pole on the Nelson Monument daily at five minutes to the appointed hour and allowed to drop at an instant which could be judged to a fraction of a second. The original wooden ball was replaced in 1855 by a more sturdy metal one. The time-ball still operates to this day—though now raised automatically—at one o'clock every afternoon.

The time-ball falling. Photographed by CPS in 1876 (RSE).

Piazzi Smyth threw himself wholeheartedly into the scientific life of Edinburgh. Immediately on being elected to the Royal Society of Edinburgh he prepared a paper on the subject of the great daylight comet of 1843 as seen by him at the Cape. He read this paper at the meeting of the Society on 6 April 1846, a meeting which for quite a different reason was an historical occasion: Piazzi Smyth's paper was followed by a short one, communicated by J D Forbes, 'On the description of oval curves and those having plurality of foci', by James Clerk Maxwell, Junior. This was Clerk Maxwell's first scientific paper. He was then a 15-year-old schoolboy at the Edinburgh Academy, and not allowed to read his paper in person, though he and his father were present to hear it delivered by Forbes. This coincidence of Piazzi Smyth and Clerk Maxwell at the same meeting, the first at the Society for both of them, clearly identifies the place of Piazzi Smyth in the historical scientific context: the one represented the traditional observational and experimental scientist, the other was destined to be the leading thinker in the revolutionary new physics.

Piazzi Smyth followed his election with several papers on technical and instrumental topics and one on the Zodiacal Light. Ever an enthusiast, he made a habit of reporting at meetings of the Society on new astronomical discoveries and communicating papers sent by colleagues abroad. He took a great interest in papers by Forbes on glaciers and by Brisbane on magnetic observations. On one occasion during his first year he presented a communication on behalf of Sir David Brewster on polarisation of light.

At that time the majority of contributions read at meetings of the Royal Society of Edinburgh were still in the old vein, giving accounts of experiments, astronomical and meteorological observations. Among the established scientists of the older generation who frequently contributed papers were Brewster, Forbes, Fox Talbot the pioneer photographer, and William Swan the spectroscopist, later Professor of Physics at the University of St Andrews. Piazzi Smyth was highly regarded by the Society from the very first; after only one year he was made a member of Council and a member of the Keith Prize Committee. In the following year he took over from Forbes, who was in poor health, as one of the Society's secretaries and was the Society's representative on the committee for the Great Exhibition of 1851. He was an assiduous Council member and only gave up the secretaryship in 1854 at the prospect of his forthcoming expedition to Tenerife. Throughout his life he remained a devoted Fellow of the Society, serving several stints on the Council and being a regular attender at meetings. The Society became his spiritual home and its rooms on the Mound with its extensive library were a favourite haunt of his for 40 years.

In 1849 William Thomson (later Lord Kelvin), then only 25 years old, read a paper in front of the Royal Society of Edinburgh on the Carnot cycle. It was the first of a wave of papers by him, his brother James, and McQuorn Rankine—all Glasgow based—on the mechanical theory of heat. This circle of brilliant young men was often heard in Edinburgh; indeed Scotland was one of the great centres of physics at this exciting period. It is interesting that Piazzi Smyth immediately sensed the importance of this new subject of what came to be called thermodynamics and that, with his inventive turn of mind, he tried to apply the new ideas to practical use. He described a gadget for 'Cooling Atmospheres in Tropical Climates', a type of refrigerator or air-conditioner which he had first pondered on in his Cape days as a result of reading in Herschel's *Treatise on Natural Philosophy* (1830) about 'the sudden expansion of condensed gases from a liquid state into vapour which is the most powerful source of cold known'. In 1850 Herschel had described an ice-making machine in an article in the *Athenaeum*, and Piazzi Smyth in a letter to Herschel explained that he had evolved a new version of his own machine: 'My claim is to determine by extensive experiments the exact amount of heat by a certain compression, comparing that with Carnot's theory and determining the most profitable compression.' A few years later he returned to the same subject in a paper 'The thermotic effect of compression of air' which was accompanied by a short communication containing the relevant calculations from an engineer friend, William Petrie, whom he had known in South Africa and who was now living in London. Piazzi Smyth made models of his apparatus and performed a series of actual experiments on compressed air at a local ironworks in which he noted the rise in temperature when the air was compressed and the fall when the compression was removed. His quantitative results were in general agreement with calculations preformed for him by Thomson and McQuorn Rankine. He claimed that with his device one horsepower could supply 20 cubic feet of air per minute, cooled 20 degrees below the surrounding temperature, a very realistic proposition. Sadly, Piazzi Smyth's lack of mathematical training did not allow him to play a part in the later development of the new physics where the young pioneers were to lead.

Piazzi Smyth occasionally attended meetings of the Royal Astronomical Society in London when he could dine with his father at the club[4] and read or communicate papers for publication. He was also elected, after some years, to the Royal Scottish Society of Arts (RSSA), a flourishing gathering of scientists and engineers whose activities were on the whole less theoretical than those of the Royal Society, though many, such as the three Stevenson brothers, the lighthouse engineers, were members of both Societies. Piazzi Smyth was President of the RSSA in the session 1858–9. He was particularly happy in this Society with its lively element of practical men and it was to this Society that he read his major papers on navigational instruments and photography, illustrating his talks with demonstrations and models. Among his early contributions was an account of a cup anemometer for use on board ship together with a set of scales and tables from which true wind speed and direction could be determined quickly from the observed apparent speed and vane direction. Another was a wind-driven device for raising water for irrigation purposes which was an extension of an actual machine which he had made as a young lad at the Cape to water his garden.

As regards astronomical problems Piazzi Smyth gave considerable thought in those early days to the question of photometric measurement of stellar brightness, a field 'barely entered on by anybody', as he said, and badly in need of being tackled. The classification of stars by magnitude was very arbitrary and only indicated that the human eye could distinguish six or seven gradations in brightness, though experienced observers claimed that they could recognise subdivisions between the magnitude steps. With the increasing size of telescopes astronomers tended to label stars down to tenth, twelfth or even sixteenth magnitude by a sort of guesswork, but the system was practically meaningless at this faintness. John Herschel, while at the Cape, tried to place stellar brightness measurement on a quantitative basis by comparing images of stars viewed directly with that of Jupiter, diminished by internal reflection in a prism. With his simple visual photometer he produced a formula relating magnitudes to apparent brightness which was useful for his purpose of estimating the relative distances of stars, assuming all to be intrinsically of the same luminosity. Herschel described his work in a paper published in 1839 and again in some detail in his *Outlines of Astronomy* which first appeared in 1849.

It was at this point that Piazzi Smyth returned to the problem which he had already discussed in a short paper while at the Cape. He considered using a wedge of tinted glass as a means of reducing the light of a comparison star to equalise it with that of the star to be studied. A major difficulty in photometric comparisons of stars, as Herschel had already noted, was the difference in their colours. To get over this problem Piazzi Smyth suggested using a small prism in the focal plane of the telescope which would produce a short spectrum and which when rotated would allow the light of the star and its comparison source to be seen in the same colour band. This would have been effectively a broad-band photometer, an idea several decades ahead of its time.

Piazzi Smyth now turned to designing a photometer which would measure stellar brightness in absolute terms, without relying on making visual comparisons between stars. Not happy with the uniformity of the wedges of tinted glass which he had tried he turned to the use of crystals of tourmaline. A pair of crossed tourmalines on being rotated could diminish light by precisely known factors; but in spite of many attempts Piazzi Smyth failed to find sufficiently good crystals. The crossed tourmalines and the wedge photometer were independently adopted later, the former most conspicuously by J C F Zöllner in Leipzig who in the 1860s invented a polarising photometer with three Nicol prisms, and the wedge photometer by E C Pickering at Harvard and Charles Pritchard in Oxford. Piazzi Smyth's prior ideas were, however, fully recognised in R Wolf's *History of Astronomy*. Piazzi Smyth's lack of a suitable telescope with which to pursue his photometric experiments was one of the reasons why he pressed for a large equatorial telescope from an early stage, though he received little sympathy at first from his Board who were convinced that the observatory's main duty was in positional astronomy.

Piazzi Smyth was ever a great lover of travel. However busy he might be he found time to make frequent trips to Hartwell House to visit his old family friend John Lee and make observations with his father's beloved equatorial of objects like Comet Encke in 1848. In 1850 he joined the British Association for the Advancement of Science at its Edinburgh meeting and attended Association meetings regularly thereafter, usually reading short papers. At the 1850 meeting he read two, one on comets and another on the work of the Edinburgh Observatory. At the Belfast meeting in 1852 he described some of his nautical instruments and mentioned his attempt to observe solar prominences outside eclipses. In this paper he declared what he was so often to repeat, the advantage of a high mountain for astronomical observation of faint phenomena. During the Christmas vacation of that same year Piazzi Smyth returned to Ireland to visit Lord Rosse's observatory at Birr (Parsonstown), and inspect the great six-foot reflector, for many years the largest telescope in the world. At the 1853 meeting of the Association in Hull, which he attended, a committee was set up to report on the physical characteristics of the Moon for comparison with geological features on the Earth. Participating observers were asked to observe various regions of the Moon under different illuminations in order to get a better picture of the shapes of its mountainous features. Piazzi Smyth joined in the project and at the Association's meeting in Liverpool the following year was one of only two observers—the other being Professor Challis of Cambridge—who had completed their part in the first year. Though conscious that his small telescope in Edinburgh was hardly adequate, he carried out his assignment of a study of 'Mare Crisium and its Shores', and produced three beautiful coloured drawings of the region which were highly praised and which were reproduced in the *Edinburgh Astronomical Observations.* It took several years to complete the British Association's map of the Moon which appeared only in 1866.

Among other journeys made by Piazzi Smyth in the first six or seven years in Edinburgh were a visit to Paris in 1849 which provided a large portfolio of water-colours; and visits to London, Oxford, Cambridge and many places in Scotland.

On 28 June 1851 there occurred a total eclipse of the Sun, visible from northern Europe. Piazzi Smyth went to Norway to observe the eclipse in the company of T R Robinson, Director of Armagh Observatory—and incidentally the inventor of the cup anemometer

Mount Gassendi region of the Moon observed in Edinburgh with Lee's six-inch equatorial telescope by CPS. Painting by CPS (RSE).

which Piazzi Smyth later adapted for navigational use. Typically, Piazzi Smyth was able to arrange the journey in a convenient and enjoyable way by engaging the interest of the Commissioners of Northern Lights in Edinburgh who offered to convey the two astronomers in their steamship to Bue Island, north of Bergen, and to assist with astronomical observations. On board on his annual tour of the northern Scottish lighthouses was the chief Commissioner, Alan Stevenson, of the famous family of engineers and the constructor of that wonderful feat of engineering, the Skerryvore Lighthouse completed in 1844. Though the sky was unfortunately cloudy at the crucial time, Piazzi Smyth, never idle, produced some beautiful paintings of the darkened landscape during the eclipse and later wrote a most instructive article on the problem of how best to reproduce a scene which lasts only a few moments. His eclipse pictures demonstrate Piazzi Smyth's remarkable gift of perfect colour memory which had been commented on already in his boyhood when he could be found indoors adding colours to outdoor sketches. Throughout his life

Robert Stevenson's lighthouse on the Island of May. Photograph by CPS (RSE).

he was often to make use of this ability to record in colour places he had visited or transient meteorological phenomena he had observed.

At this time solar physics was still at a primitive stage; there was not even general agreement on whether the solar corona or the prominences were solar, lunar or atmospheric in origin. James Nasmyth, the Scottish engineer and inventor of the steam hammer, who made a considerable reputation for himself in several fields of practical science including astronomy and photography, had suggested to his friend Piazzi Smyth that it might be possible to see solar prominences if eclipse conditions were created artifically. His method was to project a telescopic image of the Sun on to the lid of a box which had a hole of the same dimensions as the solar image. The brilliant light of the Sun's disc having been removed into the blackened interior of the box, the surrounding part of the sky might be studied without interfering glare. The arrangement was a sort of simple coronograph in reverse. Piazzi Smyth attached Nasmyth's apparatus to the small equatorial telescope on Calton Hill but failed to see any prominences. He then tried to create an artificial eclipse of the Sun by obscuring its image with metal discs or spheres which he placed on the top of the Nelson Monument on Calton Hill. When the Sun's disc was in this way momentarily obscured the edges of the 'eclipsed Sun' could be scrutinised by telescope from the ground—again without success. Piazzi Smyth rightly concluded that his failure to see prominences was due to the overpowering brightness of the sky close to the Sun. To reduce that brightness became one of the experiments in his later high-altitude observations from the island of Tenerife.

An imaginative idea of Piazzi Smyth's was his attempt to measure 'solar refraction', a possible effect suggested by William Thomson which would indicate the presence of 'luminiferous ether' around the Sun. Those were the days of the theory that light waves were carried through a substantial medium, the 'ether'. Thomson suggested that a material ether ought to have an observable refractive index which might be looked for in the vicinity of the Sun where the ether might be more dense than elsewhere. Looking through his predecessor's records, Piazzi Smyth found some good positional observations for a group of stars in the constellation of Orion and he repeated the positional measurements at a time when the Sun was among the same stars in the sky. The experiment, from the observational point of view, is of the same nature as the more recent search for the Einstein gravitational shift during solar eclipses when a small displacement of the stars' positions towards the Sun is observed. Piazzi Smyth's results were, naturally, indecisive, the tiny change recorded being within the limit of his estimated error, but it is interesting that it occurred to him to make the attempt and that he appreciated that such an experiment could only be carried out satisfactorily at the time of total solar eclipse.

A strong element in Piazzi Smyth's life was his love of art. His father published in 1851 an unusual and highly artistic book as a tribute to his friend John Lee of Hartwell who, in addition to his observatory, had in his house a magnificent museum of Egyptian antiquities and a valuable collection of Roman coins. The Admiral wrote the text, which consisted of a history of Lee's family, a description of the house—which had been the home of the French royal family in exile during the revolution—the museum and the observatory and their contents, as well as treatises on astronomy and Egyptology, all written in his distinctive enthusiastic style. The book, *Aedes Hartwellianae*, truly 'a family piece', was illustrated with the cooperation of his wife, his children and even his son-in-law Baden Powell. Piazzi's notable contribution was a coloured painting of Comet Encke, and illustrations of the transit room of the observatory and of the giant Egyptian statues in the museum.

The relation between art and science appealed strongly to Piazzi Smyth so that it is no surprise that he was one of the first members of the exclusive Edinburgh Aesthetic Club, founded in 1851 'to promote enquiry into the fixed Principles of Beauty as displayed in Nature and in Art'. Kelland, the Professor of Mathematics, was President and the membership was limited to 20. Piazzi Smyth later became Vice-President of the Club where he made the acquaintance of D R Hay,[5] the artist who decorated Abbotsford House for Sir Walter Scott, but was perhaps less well known as the friend to whom James Clerk Maxwell's father went to ask professional advice on 'James' ovals' when the boy worked on his oval curves. It was he who later supplied Clerk Maxwell with the coloured papers for his famous experiments

A social gathering in 1849, by CPS. The pensive young man, third from the left, may be Piazzi himself (RSE).

on colours and colour vision. Hay was the author of a book called *The Orthographic Beauty of the Parthenon* which Piazzi Smyth presented to the University of La Laguna on Tenerife when he went there in 1856. In short, Piazzi Smyth was immersed in every kind of intellectual activity, a prominent figure in the life of the 'Athens of the North'. 'I was delighted to see how much Smyth is respected in Edinburgh' wrote a South African friend who visited him.

Amid all these activities Piazzi Smyth found time for the lighter side of social life. He was an active member of the Symposium,[6] a dining club of Edinburgh University professors founded as early as 1792, which having fallen into apathy for some years was revitalised in 1848 with regular convivial evenings. A dinner on 22 December of that year was attended by 23 professors with the Principal in the Chair. 'After an evening spent in much hilarity, the diet was adjourned at 11 o'clock to the Observatory on the Calton Hill where Professor Piazzi Smyth enlightened his colleagues by an exhibition of the starry heavens through a powerful telescope.'

Chapter 3

Professor of Practical Astronomy

In his enthusiasm to encourage an interest in astronomy among students Piazzi Smyth decided to take up the professorial side of his post by starting a formal course in practical astronomy. He was no doubt eager to join his university colleagues as a teacher as well as a scholar. This turned out to be an unfortunate step, but at the time, after two years in office and with the completion of the reductions of Henderson's observations in sight, he saw no limit to his energy and capacity. In his childhood he had been troubled with an impediment in his speech, but this was now no more than a 'peculiar hesitation', and in the Royal Society of Edinburgh at least he was considered to be a brilliant lecturer.

To the members of his Board Piazzi Smyth argued that scientists who are supported at public expense have a duty to tell the public what they have learned, and to 'render the path more practicable for subsequent labourers', in other words, that academics have a duty to teach as well as do research, a familiar modern argument. The Board, however, was not impressed and reminded Piazzi Smyth that the Chair of Practical Astronomy was quite specifically intended to be a non-teaching one. There was no proper lecture room and no apparatus. Believing that Piazzi Smyth might perhaps be seeking to augment his salary by student fees, which in those days were paid directly to the professors, the Board urged that his salary as Astronomer Royal be increased.

However, Piazzi Smyth was not deterred, even after a university committee[1] made up of Principal Lee and Professors Forbes and Kelland recommended that, if such a course were to be introduced at all, it would have to be strictly confined to 'Practical Astronomy and Navigation' so as not to interfere with the astronomical topics already taught in the classes of mathematics and natural philosophy. These topics were spherical trigonometry, physical astronomy, optics, meteorology and 'the art of ordinary Surveying', and the committee's report laid down that these rules would have to apply not just to Piazzi Smyth but to any future Professor of Astronomy. They agreed, very reluctantly, to let the new course be tried but, until it had proved itself, the university was not prepared to support it financially apart from a rather ludicrous grant of £10 for apparatus.

Piazzi Smyth raised no objection to the proposed restrictions on the curriculum though he could not help being sarcastic about 'the magnificent sum of £10' and though he was distressed at the lack of sympathy from his two colleagues. However, had he only realised it, they were trying to save him from himself. Forbes in particular had good reason for being anxious at Piazzi Smyth's new venture: as Chairman of the Board of Visitors of the Observatory he foresaw how Piazzi Smyth's new commitment would adversely affect the work of the observatory where the warmly approved solar motion programme had a disappointingly slow start.

In spite of all discouragements Piazzi Smyth went ahead and opened his class in the winter of 1850. It was a course of 30 lectures extending over half a year with weekly practical classes in the university and weekly evening sessions at the observatory. He overcame the problem of lack of apparatus by applying to his father and to Dr Lee of Hartwell who generously lent various instruments including an eight-inch object glass which was mounted in the central dome of the observatory. Lectures were held in the military surgery classroom in the university quadrangle, and the 'globes, magnetometers and compasses' were laid out 'in glass cases out of which skeletons and models of tumours and broken legs' had been removed. The class had access to the roof of the university building where Sir John Leslie, formerly Professor of Natural Philosophy, had erected in earlier times an observing platform which could carry small telescopes and transit instruments.

The formal syllabus agreed by the university for the course on 'Practical Astronomy and Navigation' suited Piazzi Smyth very well. He had practical experience in meridian astronomy and geodesy, and through his father he had a thorough familiarity with the field of navigation. The printed syllabus and lecture notes, which survive, illuminate Piazzi Smyth's approach to science. He concentrated on conveying to his students the practical methods of making observations, the limitations and errors of instruments and the way in which observations ought to be reduced. He taught what he knew, but the course was highly specialised, and of little interest to students unless they intended to make a career in positional astronomy or navigation.

A student observing with a sextant on the roof of the university, by CPS (RSE).

Piazzi Smyth had no time for pure theory or what he called 'Utopian methods', and with such sentiments it was appropriate that the prize for the best student of the year was a copy of the *Nautical Almanac*! His efforts were, however, appreciated by the class of 12 students who turned up in the first year; he took a personal interest in them, inviting them to offer their 'free opinions' and—a century ahead of his time—allowing them to take their books into the examination room.

In addition to the formal lectures Piazzi Smyth also gave others on the history of astronomy which reveal his wide knowledge of classical antiquity and archaeology. These were the only discourses of a general nature which he ever gave. He strongly disapproved of popular scientific lecturing and writing, and in this he was at one with the members of his university committee who were critical of the

Professor of Practical Astronomy 'in another University' where 'in consequence of the Professor teaching Popular Astronomy the Practical Department had falled into abeyance'. The reference was to the University of Glasgow and to its Professor of Astronomy, John Pringle Nichol, a prolific writer of popular books and an eloquent public lecturer.

Piazzi Smyth's Inaugural Lecture—the important first lecture of the session given by every professor and intended to help students to choose their courses—is of particular interest as a statement of his own vision of the purpose of pursuing the science of astronomy. In answer to the question 'What is the use of practical astronomy?' he made use of a topical example: the discovery of the planet Neptune in 1846. The existence of another planet beyond Uranus had been suspected from the perturbations in the motion of Uranus, and its predicted position had been worked out independently by Adams in Cambridge and Le Verrier in Paris. The actual discovery of the planet, however, was made in Germany on the basis of Le Verrier's predictions. 'You will remember', Piazzi Smyth said to his class, 'the difficulty of deciding the respective claims of Adams and Le Verrier. The world seems to have decided now to give them equal praise; both have been rewarded by their respective countries and to both the public has applied the designation of great astronomers. Great they both certainly were in mathematics, great in theoretical astronomy and in physical astronomy; but yet the term "astronomer" in all its wider significance could not properly apply to them; and it cannot apply to anyone who however high his attainments in cognate sciences has yet never looked through a telescope or taken any part in the practical operations.' As for the astronomer's duty, it was, according to Piazzi Smyth, to accumulate ever more accurate positional observations for the purpose of making new, as yet unknown, discoveries in the future, following the example of the great practical astronomer Tycho Brahe: 'As Kepler, out of the smallest quantity to which he could depend in Tycho Brahe's observations elicited the elliptical motions of the planets, so, out of smaller quantities still in our day will be elicited the difference between truth and error and the roads opened up to the greatest discoveries. Had not observations been improved in accuracy since that time, Neptune would never have been discovered.' The purpose of astronomy was 'to increase the power of instruments and the accuracy of observations', even if it should be 100 or 1000 years before the results came to fruition. The practical astronomer's life was not easy; unlike speculative science 'which may either be taken up or dropped again whenever it may suit the caprice of an impulsive mind and may be prosecuted under circumstances of luxurious bodily comfort, observation has no seductive attractions about it'. The astronomer must not be tempted 'to launch his raft on an unfathomable ocean before he has a sufficient number of planks to support him'. In all scientific work Piazzi Smyth believed strongly in the value of practical skills, in 'more appreciation for the mechanical arts, more communion between the workers in metal and the workers in paper, between the labours of the hand and the labours of the head'.

Piazzi Smyth had not anticipated that his teaching load would be as heavy as it turned out to be. At the end of the first session he had to confess this to his Board, who for the first time were critical of his annual report. 'Important though the education of practical astronomers may be, this Board whose business is chiefly with watching the management of the Observatory cannot but notice a disproportionate sacrifice of time for the purpose of instructing a very limited number of pupils.' The number was indeed limited, because all but two of the first year's class were 'amateurs', meaning non-graduating students who did not take examinations, and the position did not improve in the years that followed. Piazzi Smyth's interest in university matters waned. He gradually ceased to attend Senatus meetings and after 1855, when he started planning his mountain astronomy, his name never again appeared in the minutes of that body. This did not mean, however, that he severed his links with the university; far from it. He retained his many friends there, and was a prolific user of the university library. The disappointment of the teaching course, with only an occasional student presenting himself after 1855, sowed the seeds of rancour between Piazzi Smyth and the university authorities which was to come to a bitter battle 20 years later. Had he been wise he would have withdrawn the course after the initial experiment; but it was not in his nature to give in.

In the early 1850s the possibility of setting up a large reflecting telescope in the southern hemisphere came under serious discussion among eminent British scientists who included Colonel Edward Sabine, the geodesist, T R Robinson, John Herschel and Lord Rosse. The original proposal was for a telescope of no less than three feet in diameter to be set up at the Cape, but later the specific location was left open and the suggested size was increased to four feet. In considering a suitable astronomer to put in charge of the instrument, the name of Piazzi Smyth was put forward and received general approval from the joint 'Southern Telescope Committee' of the British Association and the Royal Society. Piazzi Smyth held strongly the view that the telescope should be placed on a mountain top, and he was also determined to preserve his status in the event of being offered the appointment. The scheme however, came to nothing. Though favourably received in principle by the government in 1853, it lapsed with the outbreak of the Crimean War. The idea of a southern-hemisphere telescope was later revived, but in a different form, and led to the erection of the four-foot reflector in Melbourne in 1868.[2]

In 1855 Piazzi Smyth visited Paris to take part in the Paris Universal Exposition, an international scientific and technical exhibition inspired by the successful Great Exhibition of 1851 in London. For the Paris exhibition Piazzi Smyth prepared no fewer than 20 exhibits of instruments of astronomy and navigation, all displaying his imagination and ingenuity in producing devices for which he perceived a need. Unfortunately none of the models survives though some were described in detail in published papers read to the learned societies in Edinburgh. Piazzi Smyth's list of instruments, with acknowledgments to the local instrument makers who had constructed them, was handsomely printed by the university.

Among his nautical instruments was a 'hemispherical float log', a device for measuring the distance moved by a ship at sea which was a variation on the Massey log but used hemispherical cups instead of vanes for causing the log to spin in the water. The instrument had been inspired by discussions with the engineer Alan Stevenson who had demonstrated the shortcomings of the vane in tests in fast running Scottish rivers. Another was a ship's anemometer evolved by Piazzi Smyth in collaboration with Henry Toynbee, his brother-in-law, to record the velocity of wind at sea for meteorological investigations, in which the anemometer was placed on the mast head and furnished with electrical contacts to record continuously the velocity and direction of the wind.

Another set of exhibits, including his 'Edinburgh Universal Instrument', was concerned with the measurement of altitudes of the Moon and stars at sea. All these instruments used spirit levels whose liquids and the sizes and shapes of whose bubbles had been carefully chosen after many tests to give maximum accuracy and quickness of response. Undoubtedly Piazzi Smyth's most ambitious navigational instrument was his 'free revolver stand', a gyroscope mounting for an altazimuth telescope, intended to preserve its level on board ship independently of the ship's movements. He was able to put this instrument to the test a year later on his voyage to Tenerife.

For 'cosmical astronomy' from dry land, Piazzi Smyth exhibited his 'Edinburgh Equatorial', a multipurpose telescope designed for the use

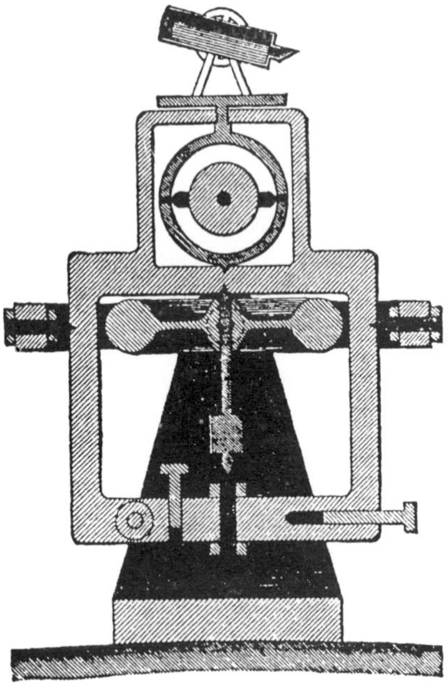

The free revolver stand (CPS, Trans. RSSA vol. 4).

of his university students. He demonstrated also special telescope mountings intended to be portable but firm for astronomical expeditions. The last in his long catalogue of exhibits was a 'Reflecting Portable Universal Equatorial' for the travelling astronomer which incorporated all the advantages of the other models, a short axis, latitude adjustment, constant eyepiece position and clockwork drive. It was mounted on a hollow stand which could be filled on the spot with water, stones or shot, Piazzi Smyth's idea of a truly portable telescope.

Commenting on Piazzi Smyth's instruments James Clerk Maxwell wrote to his cousin Henry Clerk:[3] 'Some are very ingenious, but one or two are absurd, I think. I cannot conceive a revolving wheel to be adjusted so well as not to shake its support when the velocity is great.' Yet the revolver was the instrument on which Piazzi Smyth placed the greatest store and on which he was to devote very special efforts for many years.

On his way to Paris for the Exposition Piazzi Smyth spent a rewarding day with his friends Colonel and Mrs James at the Ordnance Survey Headquarters in Southampton before sailing for Le Havre. James, later Sir Henry James, formerly regional director with divisional headquarters in Edinburgh was now the recently appointed

Director General of the Survey. His first major task there, for which he became famous, was the reduction of all previous survey maps to a standard scale, a feat which he showed could be performed with great accuracy by photographic methods. The process was demonstrated publicly for the first time at the Paris exhibition. Piazzi Smyth who had experimented with lithography and had an ambition to acquire lithographic equipment for the observatory in order to be able to copy astronomical drawings, was fascinated by what he saw in the Southampton laboratories and amazed at the extent of James' responsibility and power, with his 'hundreds of engravers and printers, and thousands of surveyors'.

It was on this tour to Paris that Piazzi Smyth began to keep regular detailed records in his journal. His account of his experiences in Paris is lively and amusing. Arriving with only ten days to go before the opening of the exhibition, Piazzi Smyth found great confusion everywhere. Only one of his three boxes had arrived and the young captain of the Royal Engineers, delegated to look after the interests of the British exhibitors, instead of attending to the waiting inventors could be seen riding in the Champs-Elysées wearing 'mustachios and

An Englishman reads his morning paper in Paris, by CPS (RSE).

extraordinary coats covered with braid'. However, all was ready in time; Piazzi Smyth's instruments were displayed in their plate-glass cases and the exhibition was duly opened by the Emperor Napoleon III on 15 May.

While waiting for the opening Piazzi had time to look around Paris. He found it much improved since his visit six years earlier, with Louis Napoleon's great streets thronged with people, the shops full and plenty of business. He visited the Louvre, strolled through the Champs-Elysées and the Tuileries Gardens and made sketches of the 'bonnes' with their babies and of the soldiers in their multitudinous uniforms. In the house of an old naval friend of his parents, Admiral Mathieu, an astronomical historian, he met the daughters of the house and their 'cousin demoiselles' and found 'the girls all rather lively; the second daughter rather pretty'. This is the only reference to the opposite sex in Piazzi Smyth's notebooks, though his earlier sketchbooks contain portraits of many charming ladies.

Before the year ended he was married.

Jessica Duncan was a highly educated and serious-minded woman who was 40 years of age, four years older than Piazzi, when she married him on Christmas Eve 1855. Her father, Thomas Duncan, died when she was still a young child, and her mother, a surgeon's daughter, married Harry Leith Lumsden of Auchendoir, a Highland 'laird' with an extensive estate near Rhynie in Aberdeenshire. On this estate of Clova Jessica spent a very happy youth. Her stepfather was an enlightened landowner who in 1825 had carved out of a barren moor a village named Lumsden for his tenants. He died in 1844 when Jessica was 29 and as he had no children of his own the estate passed to a relative. Jessica with her mother later moved to Edinburgh where she tried to improve her knowledge of geology, in which she took a special interest, having in all probability been introduced to it by Alexander Mackay, Minister of the Free Church of Scotland at Rhynie, an amateur geologist who was known to Sir Roderick Murchison and other leading Scottish geologists of the time. Geology was not taught at the University of Edinburgh until 1871, and in any case women were not admitted to the university until much later, but the subject was taught by Alexander Rose, founder of the famous Edinburgh Geological Society. Jessica was able to take Rose's courses of lectures and field work in the Edinburgh area. Her lecture notes on geology, mineralogy and mapping survive and there is no doubt that in time she became a well-qualified geologist. Her own field excursions over a period of five years took her to various parts of

Scotland and England, the Isle of Wight, Ireland, and finally, in 1854, the year before her marriage, through parts of France, Switzerland, Germany and Italy. She extended her knowledge at meetings of the British Association for the Advancement of Science where at the Belfast Meeting of 1852 she probably first met her future husband. A month later on a geological journey through England, she and her mother spent a week at St John's Lodge, the home of the Smyth parents, and visited Hartwell House and Museum. Two years later, *en route* for the Continent, she was once again the guest of the Admiral. She was the perfect companion for Piazzi, sharing his passion for science and balancing by her gentle and unassuming nature her husband's impetuous personality. She became a favourite with Piazzi's family while to Piazzi she opened a wide circle of friends and relatives in Scotland. Jessica was to become her husband's most enthusiastic assistant. Her love of scientific travel was equal to Piazzi's own and his career as a peripatetic astronomer began in earnest with his marriage.

Jessica Piazzi Smyth, née Duncan (ROE).

From the day of their marriage, Jessica took the surname of 'Piazzi Smyth', though the first of these, of course, was not a family name. In this practice Jessica followed Piazzi's sister, Henrietta, who from her marriage to Baden Powell was known, as were her children, by the double surname of Baden Powell, though the first of these was simply her husband's forename. Piazzi always signed himself C Piazzi Smyth, or Piazzi Smyth, or C P Smyth; but the middle P or Piazzi was never omitted. In Edinburgh he was known among colleagues of his own generation as Smyth, but in the course of time the double-name style which he preferred became accepted. At some stage after coming to Edinburgh Piazzi also adopted the rare pronunciation of the name Smyth, with a long vowel sound, though in his Cape days his name was pronounced simply as in 'Smith'.

Chapter 4

Tenerife, a Residence above the Clouds

At the beginning of 1851, five years after Piazzi Smyth's arrival in Edinburgh, the reduction and publication in volumes 6 to 10 of the *Edinburgh Astronomical Observations* of Henderson's unreduced work was drawing to a close. Freed from this burden, and having prepared the observatory's new long-term programme, Piazzi Smyth felt that he could at last turn to his own long-held cherished ambition: to take a telescope up to a suitable mountain site where astronomical observations could be made much more effectively than in the usual positions of observatories in or near large cities. For Piazzi Smyth the contrast between the clear skies he had experienced over the hills of South Africa and the smoky atmosphere of Edinburgh, which in those days fully deserved its by-name of 'Auld Reekie', was particularly striking. He was convinced that mountain sites provided the best conditions and he discounted the belief which had arisen from the contention of the Oxford astronomer, Manuel Johnson, on St Helena, that mountain tops were invariably cloudy. While still in South Africa Piazzi Smyth had raised this matter in a letter communicated by his father to the Royal Astronomical Society in 1845: 'A large mountain is too generally associated in idea with constant clouds and storms; but two years' experience on great elevations at this end of the colony shews decidedly that the highest position is by no means, on that account solely, liable to most mist or vexed with most wind; such effects depending more on the proximity of the sea, and the configuration of the neighbouring country.' Piazzi Smyth claimed that provided one went to sufficiently high altitudes one could rise above all cloud layers and find that 'serene and quiet air' which Isaac Newton had mentioned in his *Optiks* as the only remedy 'to take away that confusion of rays which arises from tremors in the atmosphere'.

Piazzi Smyth was the very first astronomer to advocate 'mountain astronomy' arguing its advantages in terms which are entirely familiar to astronomers in our own day. In a well-argued memorandum he wrote:

> Accuracy above all things must be the ruling idea of modern practical astronomy. This being confessed it will be found that the greatest impediment to the desired accuracy is the atmosphere, an ever present obstacle and producing, with well made modern instruments, far more untoward effects than all other sources of error whatever. Putting out of the question actual clouds preventing any view of the sky, and even not stopping to consider the effect of diffusion of general daylight—though that is a consequence of the atmosphere and very prejudicial too in eclipsing the fainter objects in the sky—yet if we only take account of the smaller undulations shown by the telescope to exist in the medium, when to the naked eye it is very clear and tranquil, we yet find them there so excessive and so lawless that seldom or never can the highest or even anything like the highest magnifying power be applied which the object glass is actually provided with and would otherwise bear with advantage.
>
> The atmosphere, then, being so determined an opponent, every effort should be made to eliminate its effects as far as possible; and this can only be accomplished by rising above its grosser parts. Such was Newton's recommendation more than a century ago. And yet, though *he* recommended it, and though many observatories have been built expressly for the purpose of procuring the most accurate observations; yet not a single one has been built in the place best calculated according to Nature and Newton for the procuring in the most perfect manner the ends for which it was really established. All our observatories, instead of being built on the highest mountains and in the clearest climates have been erected at the bottoms of the lowest valleys, hardly elevated in any useful degree above the level of the sea; and worse still they are generally immersed in the smoke of our largest towns.
>
> In justice it must be allowed that many other duties have often been demanded of observatories besides making observations, duties that compelled their proximity to the haunts of men; but even allowing for such compulsion, it is strange that men have been content to work under this excessive disadvantage and under these ungrateful difficulties. So much the more fortunate, however, for the Edinburgh Observatory, if the means should at last be afforded for its occupying this vacant but promising field for the promotion of Astronomy.

On the specific question of which mountain sites might be particularly suitable for astronomical observations Piazzi Smyth came to the conclusion that the Canary Islands would provide exceptionally favourable conditions. There was the Island of Tenerife with its Peak at an elevation of 12 500 feet where telescopes could be mounted at altitudes of 10 000 feet or more and where existed 'greater facilities of every sort than on any other known mountain'.

Piazzi Smyth presented his ideas on mountain astronomy to his Board of Visitors in August 1852, requesting their support for an experimental expedition to the Peak of Tenerife. Speaking as an astronomer working in Scotland he pointed to the additional advantage which the Canaries would have as an observing site, namely that the low geographical latitude would allow him to observe there in the summer months when night-long twilight prevents observations in the high latitude of Edinburgh. As a Professor in the University of Edinburgh where the academic year in those days lasted for only six months, from November till April, he thought that he might emulate his distinguished colleague J D Forbes who taught during the winter and devoted his summer months to the exploration of glaciers in the Alps.

The reaction of the members of the Board to Piazzi Smyth's proposal was that they 'were favourably disposed to entertain the idea' but feared that the necessary expenditure would interfere with 'other recommendations' for the use of the modest government grant to the observatory. In a letter to Sir John Herschel in 1853 Piazzi Smyth repeated his arguments for the importance of a good observing site: 'Keeping in mind your former remarks on the importance of *goodness* of observations rather than number and finding also that extreme accuracy is out of the question amongst chimney pots and finding also that the brightness of the long twilight of these northern nights effectually puts out the small stars, I have, to meet all these objections, applied for leave to take the equatorial for two months every summer into the darkness of tropical nights and into the pure and the rare atmosphere at the top of a high mountain, as the Peak of Teneriffe'.[1] These ideas were kept very much alive by him and in 1855 brought before Section A of the British Association Meeting in Glasgow.

By good fortune Piazzi Smyth's Report to his Board of Visitors for 1852 reached Robert Stephenson, the great railway engineer and the inventor of 'Stephenson's rocket', who concurred with all Piazzi Smyth's claims regarding the superiority of mountain sites, confirming them from his own experiences in the Cordilleras of South America. So impressed was he by Piazzi Smyth's suggestions that he offered him a passage to Tenerife in his beloved yacht *Titania*.

During the winter of 1855–6 Piazzi Smyth received the strong support of G B Airy, the Astronomer Royal, who had become convinced that Piazzi Smyth's ideas deserved to be tested by an expedition to Tenerife. He pressed on the government the scientific merit of the proposed experiment 'while allowing all the claims of the

Edinburgh Observatory as originating the idea, as well as the propriety of confiding its execution to Edinburgh'. Supported by the First Lord of the Admiralty, Sir Charles Wood, Piazzi Smyth's proposal was formally sanctioned on 1 May 1856. The decision was widely welcomed and praised by Lt.-Col. H James of the Ordnance Map Office as that of 'an enlightened Government which had turned its attention to the promotion of science so immediately after the conclusion of peace (following the Crimean War)'.

Once empowered to go ahead, Piazzi Smyth lost no time in making the necessary preparations for the expedition which he intended to complete within the summer and autumn of that same year. A general programme was agreed with Airy who recommended that 'it should be so lightly stated as to leave Professor Smyth in the most absolute freedom as to his general course of action'. Meanwhile the Admiralty had approached the Royal Society, the Royal Astronomical Society, the Director of the Ordnance Survey and Sir John Herschel individually asking them for suggestions. The response was overwhelming in the sense that a surfeit of proposals was put forward which covered not only observations in astronomy but studies of phenomena in physics, meteorology, geology and even botany. Apart from Airy's list there were 10 suggestions from Herschel, 11 from James and no fewer than 14 from a committee appointed by the Royal Astronomical Society. Not to be outdone, the Royal Society of London added six further suggestions. While inevitably there was a certain amount of duplication among the lists, the total added up to a quite unrealistic undertaking which went far beyond its original purpose. However, Piazzi Smyth, with his usual enthusiasm, embarked on his mission, intending to perform the essential astronomical tasks and to attempt as many of the other proposals as he could.

Piazzi Smyth's immediate problem was to collect the necessary instruments as quickly as possible. Since the Edinburgh Observatory had very little equipment most of the instruments had to be borrowed from friends and colleagues. As regards telescopes, the observatory had a 3-inch refractor donated by Sheepshanks, but Piazzi Smyth was offered the loan of a substantial equatorially mounted $7\frac{1}{4}$-inch refractor by his friend H L Pattinson, a metallurgical chemist who had his own private observatory at Gateshead near Newcastle upon Tyne. A whole range of auxiliary instruments came from various people: a Herschel actinometer for recording solar radiation from Airy, a quartz spectroscope from G G Stokes of Cambridge, a thermomultiplier for infrared observations from J P Gassiot, barometers and thermometers from John Lee of Hartwell House, a sympiesometer for measuring altitude above sea-level from the

Edinburgh instrument-maker J Adie, a heliometer for observing the angular separation of double stars from James Nasmyth and polarisation apparatus from Piazzi Smyth's brother-in-law, Baden Powell of Oxford. There was also much photographic equipment of Piazzi Smyth's own, intended for scenic as well as astronomical photographs.

The problem of transporting such a formidable array of apparatus to Tenerife was solved through Robert Stephenson's renewed offer to place his yacht *Titania* and her crew at Piazzi Smyth's disposal for the duration of the expedition. Piazzi Smyth was to be accompanied by his wife Jessica who was to act as his volunteer assistant; the expedition was destined to be the first of the couple's many joint ventures. All preparations were complete in June 1856 and with instruments and baggage stored in the *Titania* the Smyths set sail from Cowes on 24 June.

Crew of the yacht Titania. Photograph by CPS (ROE).

The rolling and pitching of the yacht during the voyage, 'the rolls amounting to 25 per minute of 10 degrees each and the pitches to 27 per minute of 3 degrees each', according to the readings of his clinometer, provided Piazzi Smyth with an opportunity to test the effectiveness of his 'free revolver stand' which was to enable an observer to keep a telescope constantly directed on a distant object in spite of any wave motion of the sea. The stand was a gyroscopic mounting which employed a rotating wheel one foot in diameter and eleven pounds in weight which was suspended and balanced in gimbal rings and put into rapid rotation by two trains of smaller wheels, acting on either side of its axis. The wheel was set and kept in motion, rotating a hundred times a second, by members of the crew of the *Titania* pulling a rope 'as only sailors can pull'. Their enthusiasm was greater than the strength of the steel driving axles, one inch in diameter, which eventually broke, but not before Piazzi Smyth had convinced himself that his device was workable and capable of keeping the sea horizon level in his telescope for several minutes at a time in spite of severe rolling and pitching.

The *Titania* arrived safely on 8 July in Santa Cruz on Tenerife where the party was received by the Spanish authorities of the island and by, amongst others, a Mr Charles Smith, a graduate of St John's College, Cambridge, and a resident of Oratava (Puerto de la Cruz) for 20 years. As with the crew of the *Titania*, the residents of Tenerife were soon infected by Piazzi Smyth's ever-present enthusiasm; though officially a one-man expedition he was never short of willing helpers. In discussions on the ultimate purpose of the expedition Smith suggested that a small party should first move to Mount Guajara at an altitude of 8900 feet.

Following this advice Piazzi Smyth commenced the ascent to Guajara on 14 July. Apart from himself and his 'intrepid wife' the party consisted of the nephew of the acting Vice-Consul, Andrew Carpenter, who was to act as interpreter, the carpenter and second mate of the *Titania* and two Spanish guides. It had been agreed that the captain of the yacht, Mr L Corke, should himself stay on board while the *Titania* was lying in Santa Cruz Roads in order to record meteorological data at sea level for later comparison with those to be obtained by Piazzi Smyth himself on the mountain. The small party was accompanied by a cavalcade of twenty mules and as many muleteers who had loaded the animals with boxes full of instruments and with a range of general equipment and provisions.

The party had a long, hot and difficult climb to the summit of Guajara on which they arrived in the evening of the same day, 14 July. The next few days were spent setting up a habitable station with a number of thick stone walls to which tents and hammocks could be

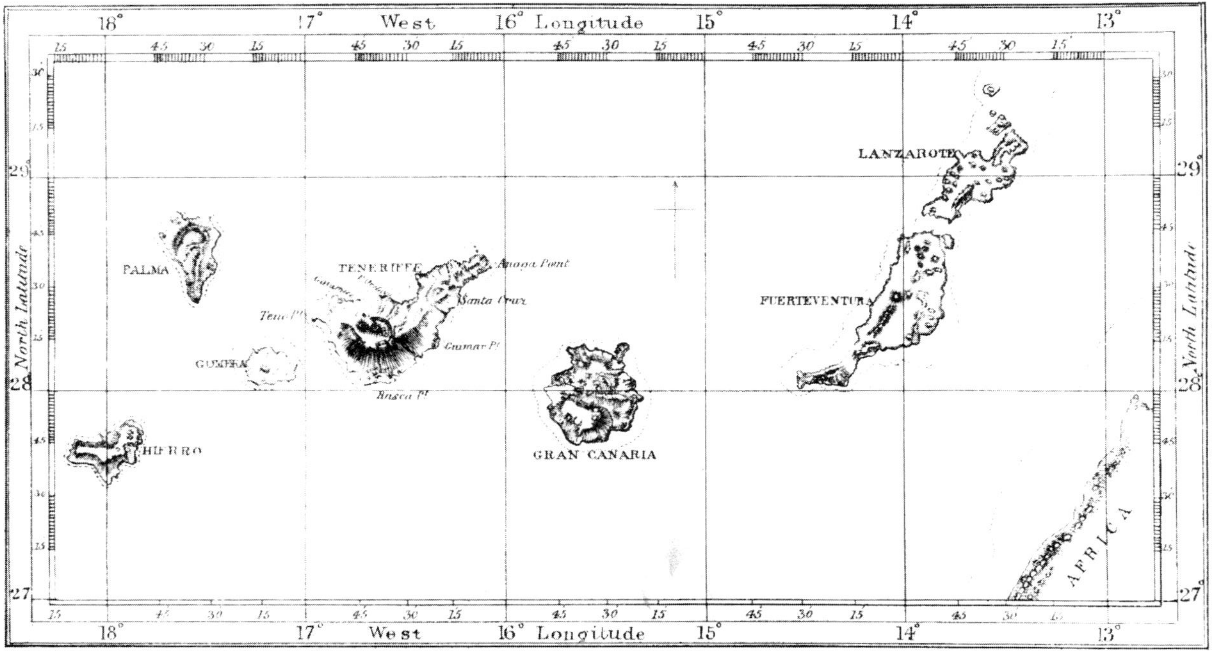

Map of the Canaries (Teneriffe Report).

Map of the Peak of Tenerife. The port of Oratava is at the top; the Guajara station is on the south rim of the crater wall and Alta Vista is close to the Peak. Drawn by CPS (Teneriffe Report).

fixed to protect the party and its instruments from wind and weather. So solid were the walls that their remains are preserved to the present time, some 130 years after the expedition.

The scientific work started immediately with regular readings of such meteorological conditions as temperature, pressure, humidity, and the velocity and direction of the wind. Within a week the Sheepshanks telescope was erected within an enclosure of stone walls five feet high. The very first observations which Piazzi Smyth made with this telescope were enough to confirm fully all the predictions which he had made on the advantages of a mountain site. He found

The Sheepshanks telescope at Guajara with the Peak of Tenerife in the background. Photograph by CPS (Teneriffe, an Astronomer's Experiment).

the clarity of sky to be such that he could readily see stars of fourteenth magnitude, four magnitudes fainter than the faintest ones which he had been able to observe with the same telescope in Edinburgh. He was greatly struck by the outstanding definition and perfect steadiness of the stellar images in his telescope which displayed very clearly the diffraction rings surrounding them, something which he had never managed to observe at home. He also found the sky so clear that he had no difficulty in picking up stars of second magnitude in daylight from one to two hours before sunset. The remarkable definition of the images of stars became particularly manifest in the easy separation of the components of close double stars. So striking were these observations that Piazzi Smyth thought it would be well worthwhile to scrutinise the double stars further at an even higher altitude. This he did later at a height of more than 10 000 feet, at Alta Vista, when he had finished his planned programme at Guajara.

At Guajara itself where the party camped for five weeks Piazzi Smyth concentrated on observations of the Sun and its spectrum, on investigations of the infrared radiation of the Moon, on observations of the shape and structure of the Zodiacal Light and on studies of various atmospheric phenomena such as the polarisation of the light of the sky. While at Guajara he also succeeded in taking a series of photographs which were all the more remarkable as they were obtained under very difficult conditions and at a time when photographic procedures were still at an early stage.

Piazzi Smyth's observations of the solar spectrum aimed at the elucidation of the origin of the dark absorption lines which had first been discovered by W H Wollaston in 1802. Twelve years later Joseph Fraunhofer had succeeded in finding more than 600 such absorption lines and in recording 324 of these in his famous map of the solar spectrum which was still the only existing one at the time of Piazzi Smyth's Tenerife observations. Using a visual prism spectroscope Piazzi Smyth found 'much general correspondence with Fraunhofer's view as to the principal lines but so great discordance as to detail that he decided to proceed on an entirely independent footing and make original drawings of what he saw'.

The spectroscope was sheltered inside a stone-walled hut roofed over with branches of trees to make a sort of darkroom, and the light was directed into a hole in the wall of the hut by means of a heliostat with a speculum mirror mounted on a pier outside. Piazzi Smyth observed the Sun over a range of different altitudes starting when the Sun was high in the sky and only twelve degrees away from the zenith, and following it until, when seen from his mountain site, it was one degree below the horizontal. The purpose of the observations being the study of the absorption of sunlight by the air of the terrestrial atmosphere, they made a major contribution to the question of the distinction between true solar absorption lines originating in the atmosphere of the Sun and others produced in the terrestrial air, a problem which for many years had been in the centre of interest of leading scientists such as Herschel, Brewster and Forbes.

Brewster and Forbes were convinced that all solar lines originated in the Earth's atmosphere. Brewster had been the first to notice, as far back as 1832, the strengthening of certain dark features in the solar spectrum when the Sun was near the horizon. During an annular eclipse of the Sun in 1836 Forbes had observed the spectrum of the Sun's edge and had found no difference between it and the spectrum of the full Sun, from which he argued that the dark lines could not be produced in an atmospheric envelope around the Sun; if that had been the case the lines would have appeared darker in the annulus where the obliquely viewed supposed solar envelope is thickest. This result, combined with Brewster's, seemed to be powerful evidence that the lines were entirely a terrestrial phenomenon. Forbes' negative result was, of course, due to the smallness and not to the absence of the effect which he tried to measure. However, his observation long remained one of the minor mysteries of astronomy, still recorded as unexplained by Forbes' son George in his *History of Astronomy* published in 1921.

In all this observational work Piazzi Smyth was helped by his unusual power of vision which allowed him to trace the spectrum into the red to a wavelength of at least 7800 Å. He had no difficulty in deciding that the feature labelled C on Fraunhofer's map (the hydrogen Hα line) was solar in origin, as it remained unchanged whatever the altitude of the Sun, while the deeper red bands, Fraunhofer's A and B, and others nearby were obviously atmospheric in origin; to use his own words, 'the lines grew visibly before my eyes' as he observed the Sun at ever lower altitudes. From his observations at Guajara he also recognised the very pronounced general improvement in the transparency of the atmosphere at the violet end of the spectrum, as he found he was able to see well beyond the spectral limit at sea-level and discover features which he had never observed before. When he repeated these observations later from the higher station of Alta Vista he found the improvement even more dramatic.

The same result demonstrating the substantial improvement in the transparency of the atmosphere on the mountain emerged from his radiation experiments. For this work Piazzi Smyth had a set of Airy's black-bulb thermometers which on exposure to direct sunlight indicated temperatures which were a measure of the intensity of the solar radiation. The increase in radiation between Guajara and the higher Alta Vista was even greater than that between Guajara and sea-level, even though the difference in altitude was considerably less—a demonstration of the effect of the atmosphere on infrared radiation, no less pronounced than that on the ultraviolet. Piazzi Smyth concluded that the black-bulb thermometer would be a useful apparatus for testing conditions on an astronomical site. He thus discovered what today is obvious but at that time was still unexpected, namely, that the termination of the solar spectrum at the red and violet ends is a consequence of the interposed Earth's atmosphere and not intrinsic in the Sun itself.

The instrument specifically designed to record solar radiation was the actinometer, invented by Herschel in 1839 in South Africa.

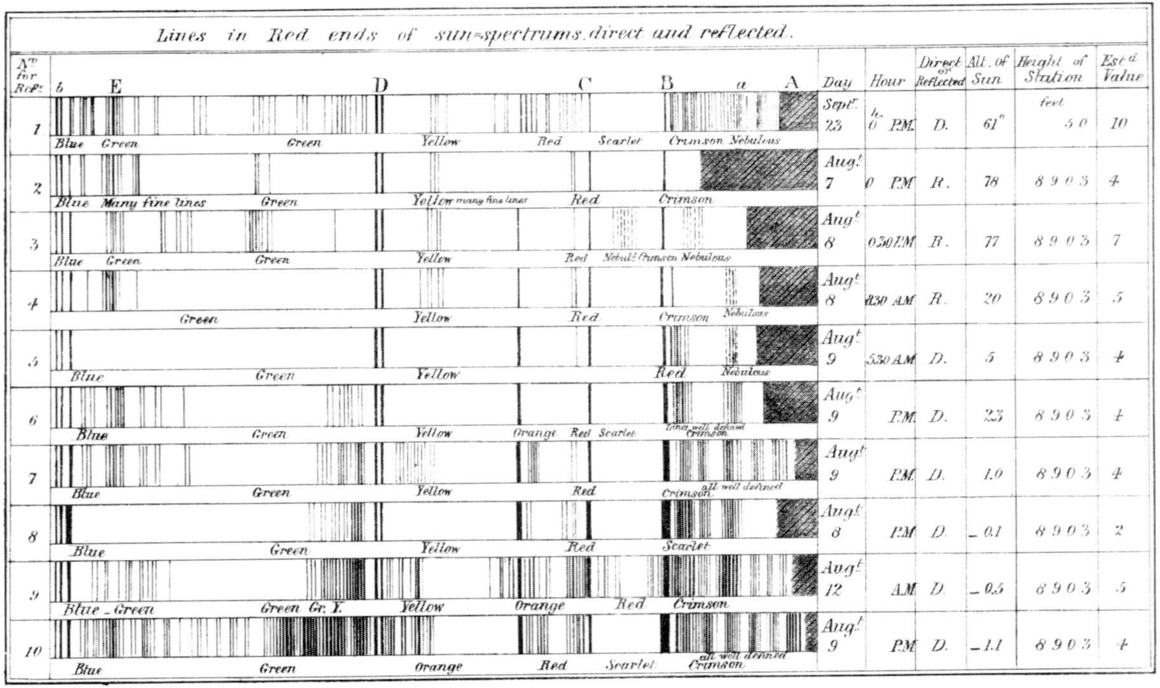

Spectra of the Sun observed at various altitudes (Teneriffe Report).

Actinometer measurements made simultaneously on the summit of the Peak and at the level of the sea for the study of the absorption of radiation by the atmosphere had been given top priority by Herschel in his list of suggestions for the expedition. Disappointingly one actinometer was damaged in transit and a second, sent out later, failed to arrive in time.

One of the experiments which had been urged by all Piazzi Smyth's advisors and which was considered of paramount importance was a search for the red prominences which had been seen on the occasions of total eclipses of the Sun as protuberances from the Sun's edge, but whose nature was still a much-disputed puzzle. Piazzi Smyth made considerable efforts to deal with the main difficulty, namely the great brightness of the Sun and the glare of the surrounding sky, using an arrangement devised by Nasmyth which he had already tried without success in Edinburgh. In this apparatus the glare of the Sun's image was removed by letting it disappear into a blackened box through a hole slightly larger than the solar disc, thus leaving on the outside of the box a projection of the surrounding sky which was viewed inside the darkroom. He did not, and as we now well know he could not, succeed in this experiment in spite of the advantage of his mountain site; but he fully realised that the failure was not due to any deficiency in his instruments but rather to the fact that the sky, even at the great altitude of Alta Vista, was still much brighter than the prominences. The problem was solved only in 1931 by Bernard Lyot with his coronograph.

A particularly important observation made by Piazzi Smyth on Guajara was his quantitative estimate of the thermal radiation of the Moon. Using his Gassiot thermomultiplier he compared the radiation received from the full Moon with that radiated by a candle placed at a distance of 14 feet 9 inches from the thermopile. In order to allow for the effect of radiation coming from the sky he pointed his thermopile alternately at the Moon and at the sky 20 degrees to the east and the west of it, just as modern radio astronomers do when they use the method of 'sky chopping'. He made the observations on two perfectly calm nights when all the other members of the party had long gone to bed and all the fires had been extinguished for several hours. To avoid the effect of his own body heat Piazzi Smyth 'swathed abundantly in non-conducting flannel' kept well away from the thermopile. The thermal radiation of the Moon was found to be rather less than that from his candle, a result not inconsistent with modern laboratory measurements. With these experiments Piazzi Smyth was the first astronomer to measure the infrared radiation from a celestial source other than the Sun.

The Zodiacal Light was brilliantly visible from Guajara where there was an unobstructed view of the horizon to the east and west. Observation of the Zodiacal Light 'with reference to recent theories' was among the suggestions made by the Astronomer Royal and by the Royal Astronomical Society. One of the theories in question was that the Zodiacal Light arose from a ring of material surrounding the Earth, in proof of which the Zodiacal Light could, it was claimed, be seen as a complete circle around the sky at midnight. The other more generally held theory was that it was a solar phenomenon, a lenticular-shaped mass of material gradually increasing in density towards the Sun as its centre and possibly composed of particles falling into the Sun and providing its energy. Piazzi Smyth, who was already familiar with the Zodiacal Light having seen it and studied it regularly in South Africa, was in no doubt about the answer even before he started, namely that the Zodiacal Light most certainly does not encircle the Earth, but spreads out from the Sun. The proof was the increase in its brightness and width near the horizon. At Guajara he watched all night when there was no Moon and found that though he could trace the Zodiacal Light to a distance of 60 degrees from the Sun, it was quite invisible at midnight. At Alta Vista he could follow it even farther out, to a distance of 84 degrees from the Sun, with maximum visibility between 3 and 4 AM but again did not see it at midnight. Airy had been right when he suggested that the problem of the location in space of the Zodiacal Light could be definitively settled on the Peak of Tenerife.

Good though viewing conditions were on Guajara, Piazzi Smyth was troubled from time to time by the presence of dust haze which he noticed up to altitudes of 9000 feet. In order to avoid that haze he determined to move to an even higher site and on the 21 August the party climbed via the 'Estancia de los Ingleses' at the altitude of 9700 feet, the usual sleeping place of travellers ascending the Peak, to Alta Vista at an altitude of 10 700 feet, the 'ne plus ultra of beasts of burden'. 'The time has now come', he wrote in his journal, 'for straining every nerve to accomplish the main purpose of the expedition, viz., to place the largest telescope on the highest available part of the mountain.' This was the Pattinson $7\frac{1}{4}$-inch telescope which was still at Oratava, as the carriage of the huge heavy boxes to Guajara had been declared impossible. Piazzi Smyth descended the mountain and with the help of a skilful German mechanic—who had also given his services as a meteorological observer—dismantled the instrument and packed its parts into seven smaller boxes which the mules could carry. The whole was carefully reassembled at Alta Vista and mounted on a pier which was a transmogrification of the telescope box, purpose-built by Cooke's according to Piazzi Smyth's design, and now weighted with stones. The reassembled telescope was housed inside an enclosure sixteen feet square with eight-foot high stone walls. With an observing chair made by the yacht's carpenter, the mountain observatory was complete.

Stereo-photograph of the Alta Vista station. Photograph by CPS (Teneriffe, an Astronomer's Experiment).

The Pattinson telescope mounted at the Alta Vista station. Photograph by CPS (Teneriffe, an Astronomer's Experiment).

The time spent on the descent and re-ascent of the mountain was not wasted; Piazzi Smyth, who could never resist an opportunity of making observations, took regular meteorological readings, placing the instruments in a 'portable observatory' fastened on a mule. This was a box made with numerous openings to allow circulation of air and covered with bright tin foil to guard against the effect of solar radiation.

The move to Alta Vista was well worth the effort. The definition of the images of stars was such that with the Pattinson telescope the components of double stars which were less than one arc second apart could easily be separated. Piazzi Smyth observed 56 double stars selected from his father's *Cycle of Celestial Objects*, measuring their separations and noting the colours of the components. His observations were then compared with other catalogues such as those of F G W Struve and Herschel.

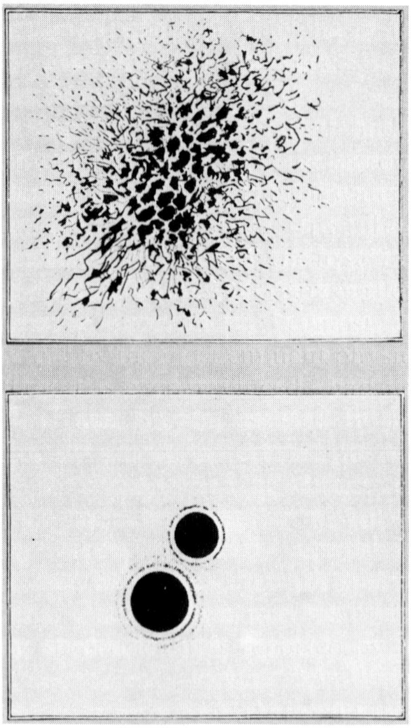

A double star seen at Edinburgh and at Alta Vista (EAO vol. 12).

On Alta Vista, too, the telescope images of planets and in particular of Jupiter near the zenith were found to be magnificent, 'a sight very different from anything that European astronomers have had for many years past'. Piazzi Smyth's drawings of Jupiter—'they might almost be called revelations'—are amongst the most impressive in his official Report on the expedition. The belts, which in the smaller telescope had 'a scollopped or festooned appearance', now showed in addition under high magnification the bright spaces between the belts resolved into 'masses of cumuli, cumuloni and cumulostrati', a description which brings to mind the beautiful photographs of Jupiter taken from the space probes of our era.

Jupiter observed from Alta Vista and drawn by CPS (Teneriffe Report).

Observations of sunspots had also been suggested, but as it was near the time of sunspot minimum this could not really be achieved. Piazzi Smyth tried instead to photograph the Sun's surface in order to show the decrease in the brightness of its disc from the centre to the border, a phenomenon well known since the early days of the telescope but never previously photographed. The able yacht's carpenter devised a camera box and a triggered shutter, but Piazzi

Smyth was not happy with his photographs and did not publish them. The 'limb darkening' was obvious to the eye on the projected image of the Sun, and Piazzi Smyth furthermore saw clearly that there was a change in colour as well as a change in brightness, the border or 'limb' being distinctly yellow while the centre was white. He confirmed this by viewing the Sun directly through the stopped-down telescope using a reflecting eyepiece to reduce the intensity of the light. In making this observation of 'limb reddening', Piazzi Smyth anticipated Angelo Secchi who is generally credited with being the first to record the phenomenon in 1870.

During the months on Tenerife Piazzi Smyth and his helpers assembled a formidable number of meteorological observations on the two mountain stations, at sea-level and on 'peripatetic excursions' around the island, when altitudes were measured with the sympiesometer, a bulky barometer-like apparatus for measuring altitude in terms of air pressure which Piazzi Smyth carried on his back. He also accumulated magnetic and polarimetric and other observations in response to the lists of experiments recommended to him, most of them of little or no interest compared with his important astronomical work.

Piazzi Smyth put a great deal of effort into recording the geological features of the Great Crater and examining ice caverns on the mountainside, reputedly below the perpetual snow level. This was done at the particular suggestion of Herschel: 'the fact should be inquired into, accurate observations made, and the concomitant circumstances carefully recorded'. Herschel's request was a tall order for an astronomer, but Piazzi Smyth exerted himself to the full and produced a long report on his findings. Even more extraordinary, coming as it did from the Royal Astronomical Society, was the proposed investigation of 'the distribution and limitation of vegetation on the moutain' which Piazzi Smyth, with his great interest in plants and flowers, obligingly undertook as well. Another botanical field of study was recommended by Herschel: 'As Mr Smyth is an expert photographer he should be provided with an apparatus for obtaining photographic impressions of everything worthy of record, inter alia, the Great Dragon Tree of Orotava, supposed to be the oldest tree in the world, from several points of view.' Piazzi Smyth thoroughly enjoyed this commission and produced several photographs in stereo of that amazing tree, 'the old patriarch of Orotava'.

The *Titania* left Tenerife on 20 September and arrived in Southampton on 14 October. If one considers the wealth of new observational material which Piazzi Smyth was able to bring back from Tenerife, his expedition, whiich lasted a total of only three months, must be looked upon as one of the most worthwhile and, with a grant of only £500, least expensive scientific missions ever undertaken.

When Piazzi Smyth's *Report on the Teneriffe Expedition* was submitted to the Royal Society of London for publication Sir John Herschel, J Challis, Plumian Professor of Astronomy at Cambridge, and Sir Charles Lyell the geologist were asked for their opinions. Herschel recommended the *Report*: 'As regards the mere question of publication by the Royal Society, there cannot be two opinions. Mr Smyth's expedition was adopted by the Admiralty as a national undertaking, embracing various important scientific objects of inquiry at a most interesting station, and the observer seems to have executed his part of the undertaking diligently and ably.' Commenting in detail on the principal answers, he continued: 'That of the advantages of a very elevated situation for astronomical observations in the way of increased efficiency at the telescope—emphatically answered in the affirmative. Jupiter drawings express details which it would be in vain to look for in any telescope at the sea-level; the definition of double stars, the tranquillity of images and the increased range in magnitude brought into view are distinct and convincing.' 'The fixed lines in the solar spectrum far more dependent on atmosphere than had been at all supposed' and 'the supposed annular form of the Zodiacal Light negative.' Herschel was disappointed that some of the instruments had got broken and could not be used but went on: 'I refer to these (causes of failure) not as casting blame but as indicating the precautions to be taken by future travellers in similar circumstances.' Herschel had only one criticism; it was to a discussion of an earlier observation by Baron von Humboldt[2] of lateral oscillations of stars seen before sunrise above the Peak of Tenerife which Piazzi Smyth had, correctly, explained as the effect of hot vapours, repudiating Humboldt's suggestion that the phenomenon might be due to lateral refraction in the atmosphere. 'The author would find it easy to do so', wrote Herschel, 'without the adoption of certain words, phrases and turns of expression.' The sarcastic tone which Piazzi Smyth was inclined to adopt with people with whom he disagreed was particularly misplaced in the case of Humboldt, an illustrious scientist and a personal friend of Herschel's.

Challis's reaction to the *Report* was the same as Herschel's. 'The part dealing with Astronomy well worthy of a place in the Transactions of the Society', and he particularly pointed to the observations of Jupiter and Saturn, the variation in the brightness of the Sun's disc between centre and border, and the Zodiacal Light. He concluded: 'The most important astronomical result of the expedition is the demonstration of the practicality of using telescopes with advantage at great elevations.' Like Herschel he found fault with the 'argumentative' parts of what ought to be a straightforward Report: 'There is much to object to in the tone of disparagement with which the observations and views of others are spoken of.'

On the geological section of the *Report*, Lyell's comments were less favourable, 'I do not think Professor Smyth enough of a geological observer nor sufficiently acquainted with the present state of the literature bearing on the theory of craters of elevation to entitle him to have this manuscript and its theoretical assumptions in the pages of the Philosophical Transactions.' Obviously Jessica's geological knowledge, on which Piazzi had depended, was inadequate, and it was no doubt rash to have tried to write a geological report in the same vein as the astronomical one where he was on firm ground. In Piazzi Smyth's defence it must be said that it was Herschel, in the spirit of the all round natural philosopher of the old school, who had instigated this geological enquiry.

Piazzi Smyth presented an account of his expedition in a paper delivered to the Royal Society on 2 June 1857. The results were published in the *Philosophical Transactions of the Royal Society* in 1858 and reprinted in a separate 138-page volume entitled *Report on the Teneriffe Astronomical Experiment of 1856 addressed to the Lords Commissioners of the Admiralty*, published at the Admiralty's expense in 1859. In the preface to this volume, Piazzi Smyth interestingly records that the astronomical chapters were published 'through the medium of the Royal Society' and the last two chapters, on geology and botany 'through the intervention of G B Airy, Esq., Astronomer Royal'. We can only guess at the manner in which Airy circumvented Lyell's objections.

The *Report to the Admiralty* was illustrated by Piazzi Smyth's drawings of Jupiter and the Moon, his maps of the solar spectrum, landscape illustrations of the cloud horizon as seen from Mount Guajara and numerous diagrams. Most striking, perhaps, was the enlarged photograph of the Alta Vista observing station. This was one of a number taken by Piazzi Smyth and, remarkably, printed by Jessica who assisted Piazzi at his photography as in all his other scientific activities. The front page of the Report was decorated by a pair of stereophotographs of a model of Tenerife prepared to Piazzi Smyth's data by James Nasmyth along the lines of his well-known popular models of lunar craters. This too was printed by Jessica. In an explanatory note Piazzi Smyth discussed the problem of producing photographic illustrations for books and publications. Bearing in mind the burst of enthusiasm which had greeted the discovery of photography and the predictions that photographs would soon replace drawings for scientific purposes, he complained at the difficulty of getting his photographs professionally printed in large numbers. 'Where is the fault or the difficulty?' he asked 'It is not in photography itself; for, as a single specimen of what might have been done in this case, my wife has, in the course of a very short space of time, printed off 350 copies of an enlarged negative of one of the Alta Vista photographs, and as many of a stereoscopic view of a model of Teneriffe kindly prepared to our data by James Nasmyth, Esq., C.E..' These unique photographs, produced by the husband and wife team, are inscribed on their mounts 'C.P.S. Phot, J.P.S. Pr.'

At the request of the Board of Visitors full details of the astronomical observations made in Tenerife and the scientific journal were included in volume 12 of the *Edinburgh Astronomical Observations* which appeared in 1863. This report, free of the extraneous forays into fields of science which were not his own, is an outstanding piece of work. Here his original purpose is clearly carried out, with tests for steadiness and transparency of the atmosphere and an investigation into the atmosphere's selective effects both on the continuous and the line spectrum of celestial bodies. He demonstrated without doubt the advantages of his high-altitude sites and showed that cloudlessness and height alone were not enough. The report in the *Edinburgh Observations* was illustrated by only a few photographs, the Stationery Office which produced it being constrained by the cost of having them photographically reproduced. In this report the photographs had an important scientific purpose, in that they demonstrated the clarity and stillness of the atmosphere at a height of 10 700 feet by the manner in which extreme detail in the distant landscape could be resolved. One of the photographs was reproduced by the method of photoglyphing, recently invented by Fox Talbot, one of the pioneers in photography. Fox Talbot was a friend of Piazzi Smyth's and a Fellow of the Royal Society of Edinburgh, who on hearing of Piazzi Smyth's desire to publish some of his Tenerife photographs offered to engrave them by his photoglyphing process, a method of copying the photograph on a steel plate and printing it in the usual way with printer's ink.

Piazzi Smyth's popular account of his and his wife's experiences, called *An Astronomer's Experiment or Specialties of a Residence above the Clouds*, appeared before either of the scientific reports. This charming book, written in Piazzi Smyth's inimitable style, was illustrated with a large number of his own very remarkable photo-stereographs, the very first book to be illustrated in this way. Piazzi Smyth was successful in getting these photographs reproduced by the publisher, Lovell Reeve of London, who included 20 stereoscopic views in the volume for the large edition of 2000 copies.

The success of the Tenerife expedition earned Piazzi Smyth the Fellowship of the Royal Society to which he was elected in June 1857 on the recommendation of a long list of Fellows headed by J F W Herschel and including Airy, Babbage, Baden Powell, Beaufort, Manuel Johnson, Makdougal Brisbane—and Lyell. His nomination was prompted by his ever-devoted parents. W H Smyth had written to Herschel suggesting that his son was desirous of becoming a Fellow. Herschel was very willing to support him: 'Your son is a very fitting person to be a member of the RS and indeed I thought he had been so.' This was not enough for Smyth who touchingly insisted that Herschel's name should head the list 'for auld lang syne'. The testimonial described Piazzi Smyth as 'The Inventor or Improver of several instruments in Navigation, Surveying and Practical Astronomy. Distinguished for his acquaintance with the science of Practical Mathematics and Instrumental Knowledge. Eminent as a promoter of various branches of abstract and applied science, and as having performed a scientific journey to the Peak of Teneriffe in 1856.'

The island of La Palma seen above the cloud layer, drawing by CPS (Teneriffe Report).

More than a hundred years later Piazzi Smyth's successors in Edinburgh were to retrace his footsteps up the Peak of Tenerife and endorse his findings. Their final choice of a site for the new generation of British telescopes was Tenerife's sister island of La Palma, which Piazzi Smyth could see rising above the layers of cloud and dust, when he looked across from his Mountain Observatory on Alta Vista.

Chapter 5

Visit to Pulkova and Three Cities in Russia

The Nicholas Central Observatory at Pulkova near St Petersburg was regarded in the mid-nineteenth century as one of the world's greatest astronomical institutions. Airy, the Astronomer Royal, who visited it in 1847, declared that every astronomer ought to visit it once in his life—just as a visit to the astronomical establishments in California was considered *de rigueur* in more recent times. A journey to Pulkova (now called Pulkovo) was Piazzi Smyth's next great desire.

Pulkova Observatory had been founded by the Emperor Nicholas I in 1835 on the advice of F G W Struve. Struve, born in Altona near Hamburg in Germany, had, like many of his compatriots, moved into the Russian Empire where he had made his mark as Professor of Astronomy and Director of the Observatory at Dorpat in Estonia. There he had devoted himself for ten years to the observation of double stars. The work led to the publication in 1836 in St Petersburg of his *Mensurae Micrometricae* which gave the places, angles of position, separations, colours and relative brightnesses of 3112 double and multiple stars, providing an invaluable standard of reference for generations of double-star observers. As a result of the reputation which Struve gained from his outstanding work he was summoned by the Russian Emperor to superintend the establishment of the new observatory at Pulkova which was to be equipped with the best instruments available. When completed in 1839 Pulkova did indeed surpass all other observatories 'in splendour, efficiency and completeness', to use Agnes Clerke's description.

In the matter of efficiency, Struve strongly believed in the principle that as long as the general scientific purpose of the observatory was clearly identified and the necessary practical equipment provided, his assistants should be left as free as possible in the pursuit of their work.

As he put it, he did not want 'mechanical observers' but 'thoughtful and creative colleagues'; to sustain happy personal relations between himself and all members of his large staff was an important part of his doctrine. At the time of the Smyths' planned visit Struve, or von Struve as he was formally called in Russia, was away in southern Europe recovering from a spell of ill-health which had compelled him to hand over most of the responsibility for the observatory to his son Otto, a man of about the same age as Piazzi Smyth and, like the elder Struve, personally known to Piazzi's father.

The question of travel had been solved for Piazzi Smyth, as on his earlier journeys, by one of his many friends. On this occasion his patron was R M Smith who was also a member of the Board of Visitors of the Observatory. He was a ship-owner who traded extensively with Russia and who was able to arrange for the two Smyths to travel—free of charge—on the modern iron screw steamer *Edinburgh* which did not normally have accommodation for private passengers. The *Edinburgh* had plenty of room on deck where Piazzi Smyth could set up his 'free revolver stand' for a test in open sea of a second improved version of the instrument which had been first tried on the voyage to Tenerife. The ship's Captain, David Steele, and his crew were quite as delighted to assist Piazzi Smyth with his experiments as the sailors on board the *Titania* had been three years earlier.

The Smyths boarded the *Edinburgh* on 29 July 1859, at the harbour of Leith, a few miles north of their home in the city, Piazzi having left detailed instructions with his assistant Wallace on how to conduct the business of the observatory in his two months' absence. They carried aboard with them, apart from seven pieces of luggage, ten heavy boxes, five of them filled with parts for the revolver experiment and five with photographic equipment. Piazzi had his notebooks to record the journey and make sketches.

They sailed out of the Firth of Forth into the North Sea and three days of stormy weather. Piazzi philosophised on the virtues of the intrepid Vikings of old who braved the gales as he himself now did while he tried out his revolver stand on the deck. He was somewhat disappointed when the apparatus did not live up to expectations in the rough seas, but he did not allow himself to be disheartened. Conditions became easier on the third day when the *Edinburgh* entered the Kattegat and passed Elsinore when Piazzi Smyth could not help reflecting on Shakespeare's story of Hamlet, and passed the Island of Hveen with its ruins, though too distant to see, of Tycho Brahe's Uraniborg, the great observatory of the sixteenth century. Piazzi Smyth liked to model himself on Tycho Brahe, and adopted Tycho's philosophy that improved instruments for more accurate observations was the way forward to fundamental scientific discoveries.

Past Copenhagen the steamer reached the Baltic Sea and sailed on to the Gulf of Finland. Fascinated as always by meteorological phenomena Piazzi Smyth recorded the unusual colours of the clouds hovering over the Russian mainland which they reached, after a voyage of five days, at the busy port of Kronstadt in the bay outside St Petersburg. There to receive them was the librarian of the Pulkova Observatory who guided them past the customs officials and the soldiery of the fortress. They were transported by droshky to the ferry for St Petersburg, arriving at last at the observatory, a few miles beyond the city.

The Smyths were received with typical Russian hospitality by Otto Struve who spared no pains to make their visit memorable. On the first evening, true to the Struves' ideas about staff relations, the whole observatory community and their families, children included, foregathered to celebrate the arrival of the visitors from Scotland. At table too were the 'optical and mechanical artists' of the observatory and a group of members of its geodetic department who had just returned from surveying expeditions, some from Odessa in southern Russia, others from Astrakhan, a thousand miles away near the Caspian Sea.

Pulkova Observatory in 1859, photograph by CPS (RSE).

With his artist's eye Piazzi Smyth admired the location of Pulkova on a hill at the edge of a deep forest. He was much impressed with the spaciousness of the observatory building, an immense contrast to his own confined premises on Calton Hill. The interior was gleamingly clean and in the extensive library, centrally heated and dust-free, the books could be housed on open shelves. When the Smyths were shown the cool subterranean caves in the basement where several months' supplies of food could be stored, Piazzi could not but wonder 'what Napoleon Buonaparte could have been thinking about to lead 500,000 men into Russia without making more preparations for their winter subsistence'. At the entrance to the magnificent main building Piazzi Smyth noticed the date of the observatory's foundation, 1839, engraved in golden figures, and also an icon of St Nicholas, its protector. At this Struve assured him that 'we ourselves do not use it, as we are mostly Lutherans', an illustration of the dominant role of German scientists in the Russia of the time when the Academy of Sciences was often criticised for keeping its doors shut to Russians while opening them too widely to German scholars.

Piazzi Smyth's first interest was naturally in the scientific equipment of the observatory which, being geared to positional astronomy, included a number of substantial meridian instruments, constructed by the distinguished German makers, Ertel of Munich and Repsold of Hamburg. There was also a heliometer by Merz in Munich, an instrument of a type which had been originally designed for the measurement of seasonal variations in the Sun's angular diameter but which was used with great success by Bessel and others for the precise measurement of angular distances between stars. However, pride of place in Pulkova was given to a 15-inch Merz refractor with a focal length of 23 feet which was particularly envied by Piazzi Smyth who was still without a suitable equatorial telescope for Edinburgh. The Pulkova refractor had been used constantly since 1841 by Otto Struve for the observation of double stars, his father's favourite field of work. The Smyths were excited at the opportunity of looking at double stars through this large instrument but were not entirely unprepared to find that the telescope's great light-gathering power was largely negated by the poor seeing conditions on that night in Pulkova. 'What a scene!' exclaimed Piazzi, 'a sort of medusa's head of twisting twirling filaments of light.' The double star so described had an angular separation of 0.3 arc seconds and had been clearly resolved with his little three-inch telescope at Alta Vista, providing a most obvious demonstration of the vital advantages of a mountain site.

Otto Struve was well aware of the importance of 'Mountain Astronomy', a term which, according to Piazzi Smyth, Struve was the first to coin. He had already studied the *Teneriffe Report* and examined the quality of photographs of distant views taken by Piazzi Smyth on Tenerife which demonstrated the clarity of the atmosphere at high altitude. Copies of these photographs had earlier been dispatched to Pulkova on one of his merchant vessels by the same ship-owner friend who had arranged the Smyths' passage. Struve told Piazzi Smyth that the Russians intended to take up mountain astronomy, adding that 'he did not wish to take an unjust advantage of him and to get there before him in the race'. The Russians, however, did not have an ocean mountain top within their borders. 'Our path is through the centre of continents and she [Russia] has now, in the extreme southern line of her territories, mountains quite high enough for these purposes without going anywhere else into foreign countries.' The reference was to Transcaucasia 'where we can go pretty nearly three and a quarter miles vertically in the air, or equal to a mile above the Peak of Tenerife'. The Russian government was at the time founding an observatory at Tiflis (Tbilisi) in Georgia as a base for geodetic operations in the Caucasian provinces which would also allow astronomers to carry their instruments up to the surrounding mountains. The speed with which the Russians could implement their plans for mountain stations was, so Otto Struve explained, because of the Emperor's personal interest in astronomy and the elder Struve's direct access to him.

In Russia, the practical task of reaching high mountains for scientific operations did not present a serious problem. While Piazzi Smyth was there a party of a hundred men had just returned from three months' surveying on Mount Ararat on the very summit of which they had placed a theodolite at an altitude of 17 000 feet above the Black Sea. Through Struve's intervention Piazzi Smyth met the Surveyor General for the Caucasus, General Chodzko, who showed him the volumes containing his original measurements, some of his survey maps and also the Scientific Report of the great triangulation centred on Mount Ararat which had just been printed. The Report was written in Russian and Piazzi Smyth bemoaned the fact that like many other Russian scientific advances it would not reach the scientific community in the rest of Europe on account of their ignorance of the language. He was enthusiastic enough to suggest that Russian might be included in the curriculum of the University of Edinburgh, which at that time was debating the question of teaching foreign languages.

One of the most impressive meetings which Piazzi Smyth had in Russia was with the Grand Duke Constantine Nicholaievich, the brother of the Emperor, who as Lord High Admiral had taken a particular interest in the operations of the Pulkova Observatory. He had expressed a wish to see the free revolver stand which Piazzi Smyth was requested to demonstrate at his Palace of Strelna, a few miles outside St Petersburg and overlooking the Bay of Kronstadt. Struve had arranged for the transport of Piazzi's instrument from Pulkova to Strelna in a 'telega' carefully packed in straw and accompanied by two soldiers who had been well drilled in working the wheels of the gyroscope. On reaching the gates of the Palace grounds the soldiers donned brilliant green uniforms with golden braid and the party entered in procession the flower gardens ornamented with bronze and gilt statues. Inside the Palace the revolver stand was set up as on board ship and demonstrated to the Grand Duke, his sons and two of his admirals, while Piazzi Smyth explained the principle of his invention. 'Whereupon', Piazzi Smyth records, 'the faithful veterans of Pulkova had the honour of exhibiting in the presence of one so near their own Tsar how they could first set the two internal revolver wheels spinning at the rate of one hundred and fifty revolutions in a second; and then give the whole mounting of the machine certain rolling and rocking movements; while an observer at the telescope indicated how steadily all the time a distant tree on the Finland coast was remaining in the field of view, and how tolerably quiescent were the bubbles of the several levels.'

The apparatus though admitted by Piazzi Smyth to be as yet imperfect worked well enough at Strelna to impress the Grand Duke who declared it 'the most promising thing we had yet seen to supply a real desideratum in nautical astronomy, enabling us to get the astronomical position of a ship as accurately by night as by day, and compete in precision with scientific travellers on land'. At this memorable meeting there was no need for the usual interpreter; the Duke spoke perfect English to the Smyths while his chosen language of communication with Struve was German. The outcome of the demonstration of the revolver stand was that the Grand Duke ordered a modified copy of Piazzi Smyth's machine to be constructed without delay using the best materials available. The 'artists' might take their time provided they produced a good instrument at the end.

Another, less elaborate, invention in which Piazzi Smyth was interested at this time, and on which he found Otto Struve was also working, was a 'distance measurer' or range finder. The usefulness of such a potential device for gunnery had become obvious in the Crimean War (1853–6), and Piazzi Smyth was one of those who put their minds to designing such an instrument. Piazzi Smyth's device had reached the stage of a prototype being made by Cooke's of York, but he was not satisfied with its performance and gave up working on it. The reason may have been that Henry Clerk, his friend from South African days and now a colonel in the Royal Artillery at Woolwich, had in 1858 brought out his own successful version of a range-finder which was similar to Piazzi Smyth's.[1] Clerk kept in touch with Piazzi on the progress of the later improvements of his distance measurer. The subject was of sufficient interest, however, that on his return from Russia Piazzi Smyth was invited to read a paper on 'Distance measuring in Russia' to the Institute of Civil Engineers in Glasgow which was published in their *Transactions*.

Though their prime interest on the trip to Russia was to visit Pulkova, the Smyths intended to see as much as they could of the country and of Russian life in general. Believing always in independence, Piazzi Smyth had armed himself with dictionaries and had studied accounts of former travels through Russia by scientists such as Granville, Murchison and Lyell. He familiarised himself with Russian history and Russian poetry through Bowring's popular translations. The plan to pursue their travels without interpreters and avoiding the standard tourist route worked well. The Smyths soon found that, when they could make conversation, the subject of astronomy never failed to make friends. Jessica bought a samovar so that they could reciprocate invitations to tea in their hotel rooms. They avoided English-speaking establishments and took their meals in Russian restaurants where they sampled local dishes and enjoyed the taste of cabbage soup and Russian bread.

On Sundays the Smyths, though believers in the simplicity of the services in the Church of Scotland, attended Orthodox Churches and felt deeply moved by the piety of the Russian congregations. Describing their experiences afterwards Piazzi Smyth wrote vivid and sympathetic accounts of baptism and wedding services which showed none of the disapproval of images, candles, holy water and processions which he displayed in later life. Piazzi Smyth wrote an interesting article on his impressions of the Russian Church for the evangelical journal *Good Words*.

After St Petersburg it was Moscow's turn to be seen. They went to the Kremlin with the Uspenskij and its other magnificent cathedrals and its giant gun and bell, to the fantastic Prokrosky Cathedral in the middle of Red Square, to various convents and monasteries—and, last not least, to the cemeteries which at all times had a strange fascination for Piazzi Smyth. Astronomical visits in Moscow were more social than scientific, but the Smyths looked at the new dome which was being erected for the Moscow Observatory on Sparrow Hills, 'the place from which Napoleon had his first and last view of the ancient metropolis'.

With a few days to spare before the *Edinburgh* was due at Kronstadt to sail back to Scotland, the Smyths adventurously travelled up the river Volkhoff to Novgorod, a city which in medieval times had played a leading part in the history of Russia. There they saw ancient churches, monasteries and convents, and also genuine Russian steam-baths where the bathers beat themselves with twigs, a long-standing health-giving custom the origin of which Piazzi Smyth investigated with his usual avidity for knowledge.

On all these peregrinations in three cities Piazzi Smyth carried his photographic equipment in many bulky boxes of optical and chemical

The Kremlin and its cathedrals, photograph by CPS (RSE).

materials. It had taken some time to obtain customs clearance for these boxes at Kronstadt at the beginning of the tour, and permission had to be requested—and was granted—from the Governor General of Moscow to take photographs in public places. The cabmen at the railway station in Moscow had demanded an extra fee for transporting such unusual baggage to the hotel. The Smyths were undeterred, however, and the labour of carrying their equipment on trains, boats and carriages through hundreds of miles bore fruit. They obtained a beautiful set of photographs, mainly of buildings and static scenes, some in stereo, of Moscow, St Petersburg and Novgorod. There were also photographs of people, but exposure times were too long to allow more than a small number of these to be taken. The photographs were later exhibited at gatherings in Scotland. Before finally leaving Russia

The Great Gun of the Kremlin, photograph by CPS (RSE).

the Smyths were privileged to be included in a family party which went to meet the elder Struve at the quayside on his return from his convalescence in Italy. At a farewell evening in his company at Pulkova Wilhelm Struve toasted his old friend, Piazzi's 'respected father'. The Smyths had been popular guests at the observatory; when David Gill visited Pulkova many years later they were affectionately remembered by the staff and their families and by the Struves' housekeeper with whom Piazzi used to converse in mime.

On the return journey to Leith, Piazzi Smyth was delighted to find that the crew of the *Edinburgh* had taken enthusiastically to the scientific lessons he had taught them on the outward journey, maintaining meteorological records, logs of air and water temperatures and the like. As a result Captain Steele joined a naval meteorological observers' group which had been set up by Admiral Fitzroy, remembered chiefly as the commander of the *Beagle* which had taken Charles Darwin on his famous surveying expedition in the early 1830s. It was clear that the Smyths had made the sea voyage as memorable for Captain Steele and his crew as for themselves.

A year later, almost to the day, the *Edinburgh* was tragically lost with all hands in the North Sea. The Smyths had spent six weeks preparing a set of photographs of their Russian tour and had taken the boxes personally to Leith harbour to be conveyed by Captain Steele to Kronstadt for their friends in Pulkova. At the time of sailing, on 2 October 1860, a strong wind was coming up which according to the customs clerk at Leith 'would send the Captain quickly across the

North Sea'. The wind developed into a hurricane which caused enormous devastation all over Scotland, inland as well as on the sea, with grave loss of life and property. Five steamers including the *Edinburgh* and many fishing vessels were lost without trace. The Smyths were profoundly distressed by the tragedy, and Piazzi wrote a moving account in his diary: '22 widows and 60 orphans. Lost! Lost!'

The storm had developed without warning and the feeling was widely expressed in Scotland that the course of this terrible event should be recorded. In the absence of any official body which would do this, Piazzi Smyth undertook the task, partly because of his strong personal interest, partly also because he believed that more ought to be done by the Meteorological Office in the field of storm forecasting. He saw the task through though it was several years before his report, *The Great Hyperborean Storm*, was completed and published.

As he had done after his return from the expedition to Tenerife, Piazzi Smyth followed the tour of Russia by an equally popular book on his and Jessica's adventures. The two volumes of this book, *Three Cities in Russia*, were published in 1862. Written in Piazzi's characteristic style and illustrated by his own drawings the book gives a delightful account of Russian life in those far-away days. The book also helped to keep the subject of mountain astronomy before the general public. Piazzi Smyth hoped that in pointing out the Russians' interest his own ambitions of a mountain observatory might be advanced. The members of his Board of Visitors were entirely favourable to the scheme, but they were powerless in face of the Treasury's refusal to grant the necessary financial support. Following the publication of the Report on the successful expedition to Tenerife the Board of Visitors had renewed their application, without success, for a grant of £3000 for an equatorial telescope; it was to be fairly large but portable so that it could be used both in Edinburgh and at a favoured site abroad. This latter condition prevented the acceptance by the Board of the offer of a nine-inch telescope which had belonged to Sir Thomas Makdougal Brisbane, a firm supporter of the observatory on Calton Hill since Henderson's time and a good personal friend of Piazzi Smyth. When after Sir Thomas' death his widow offered the telescope in accordance with her late husband's wishes it was on condition that it should be mounted on Calton Hill, a condition which led the members of the Board to decline the offer with some reluctance on the grounds that 'the telescope was not of sufficient size to be suitable for the purposes of a public observatory and that they would not feel justified in recommending to Government to incur the expense necessary for its erection'. Piazzi Smyth himself realised that to accept the Brisbane offer would jeopardise his chances of getting a larger portable instrument with which he would commute between Edinburgh and Tenerife.

Piazzi Smyth pressed his case for the implementation of his Tenerife scheme by describing the interest also shown by the French in his Tenerife Report. 'When the whole of the trial had been concluded', he said in 1857, 'an eminent French savant wrote that the little expedition which returned to England in October 1856 had inaugurated a new and powerful system of astronomical observation; and he proposed that France should at once follow in the same line, with a station on the Pic-du-Midi.' Though the idea of erecting a building 500 metres below the Pic had first been discussed in 1828, and though a mountain inn was built in 1852, the plan to set up a station for scientific work did not take hold until 1865 and the only recorded foreign contact is a letter from Sir John Herschel dated 1867.[2] It was only in 1882 that the possibility of having an astronomical observatory there was first discussed. The Pic-du-Midi became one of the most successful mountain observatories in the world but its choice as a site does not appear to owe anything to the Tenerife experiment. The French friend who made the suggestion to Piazzi Smyth in 1857 remains a mystery.

At about this time the Treasury gave a grant of £1000 towards the setting up of a hill observatory in India, a scheme which was simpler and less expensive than Piazzi Smyth's. Like Russia, the British Empire had easy access to high mountains and had scientifically trained men in their garrisons. The idea of setting up a hill station in India for a limited number of years was first brought up in 1858, the year of publication of the Tenerife Report, but had to be shelved following the Indian Mutiny. The plan was revived in 1861 by R C Carrington and W S Jacob, two astronomers who were on very friendly terms with Piazzi Smyth so that their scheme was in no way seen as a rival to the Edinburgh one. Carrington had in fact been on the Royal Society committee which advised on Piazzi Smyth's Tenerife expedition. He was a Cambridge-educated scientist who after some years as director of the University Observatory at Durham set up his own private observatory in Surrey where he did most of his work first on star catalogues and later on sunspots. Jacob, an army engineer with the rank of captain, began his scientific life as a surveyor in India, where he set up his own private observatory in Poona, near Bombay. He was later appointed Director of the Observatory of Madras where he did some good work on parallaxes and double stars. Piazzi Smyth reported some of these observations at the Royal Society of Edinburgh.

Jacob put his proposal of a hill observatory near Poona, 4000 feet above sea level, to the Royal Astronomical Society, which recommended it to the Government.[3] The grant was provided to set up a station for three years with a nine-inch telescope which Jacob himself had purchased. Piazzi Smyth welcomed the project and Jacob, within a few days of his arrival in Bombay, wrote to Piazzi Smyth to say he was about to set out for the mountain. Five days later he died suddenly in Poona.

Piazzi Smyth's mountain astronomy was different from Jacob's in that he wanted a 'peripatetic observatory' for Edinburgh at the proven best possible site, and not merely a station on a convenient elevation of a few thousand feet. The difference between his contemplated observatory and stations such as the proposed Indian one was well set out by Jessica in a letter to John Lee,[4] who wrote in connection with the Jacob expedition which he had supported: 'My husband and myself have noticed only in the public papers with great pleasure the admirable and successful exertions you have made to secure assistance to Captain Jacob, for though his idea cannot be called a carrying out of the 'Astronomer's Experiment', still his efforts are in the right direction and his work is sure to be first rate.' In the same letter Jessica told Lee of Piazzi's forthcoming popular illustrated article on his Tenerife experiment for *Good Words*: 'He was asked by the Editor and Publishers of the Magazine to write an article for them to suit the title 'Above the Clouds', clearly showing that the subject is at last beginning to be popular in the community, seeing that the publishers have to cater monthly for the circulation of 50,000 copies. I suppose we need not expect Government to establish the system until the pressure comes from without—but if that did come, and also the Russians prove successful with their noble Ararat expedition, we might hope to see a summer astronomical observatory established on the Peak of Teneriffe!' Soon afterwards Lee informed the Smyths that the Tenerife expedition was to be commemorated on the Moon by 'Crater Piazzi Smyth', the 'Pico' itself and the 'Teneriffe Mountains'. These features were incorporated in Edmund Neison's Lunar Atlas published in 1876.[5]

Otto Struve's plan for mountain astronomy went ahead in the meantime. In 1861 a Russian site-testing expedition—referred to in Jessica's letter to Lee—set out for Mount Elbruz in the Caucasus at an altitude of 18 000 feet, and in 1863, four years after Piazzi Smyth's visit to Pulkova, an observatory was opened at Tiflis, both to assist geodetic observations and to operate as a first step in the Russian plan

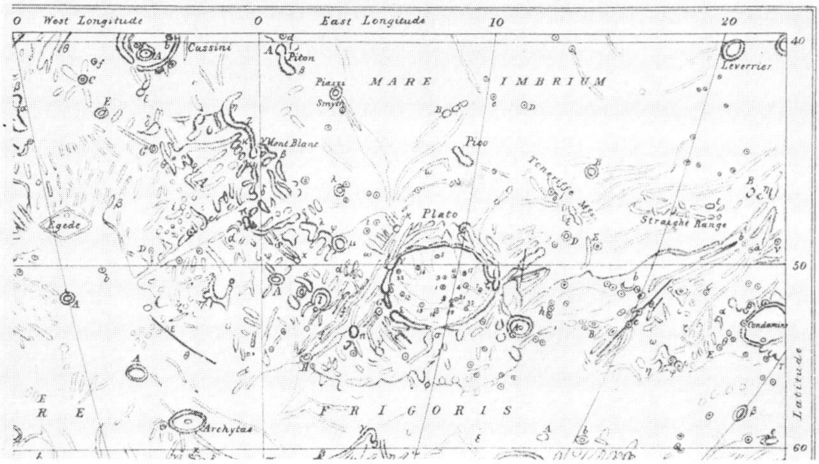

Crater Piazzi Smyth, Pico and the 'Teneriffe' Mountains on the Moon (Neison, The Moon).

for a high-altitude observatory. Such an observatory was in fact established in 1892 at Abastumani in Georgia at an altitude of 4600 feet, much lower than the Peak of Tenerife. The scheme took considerably longer to realise than Piazzi Smyth had anticipated when he looked to Russian progress to speed up his own plans. It was overtaken by the Lick Observatory set up on Mount Hamilton in California in 1876.

The idea of the free revolver stand which had caught the imagination of the Grand Duke Constantine Nicholaievich, was pursued further in Russia and in Scotland. The Russians worked for about two years on prototypes, without much success. Piazzi Smyth acquired new mechanical parts from Milne's and spirit levels with special large bubbles from Adie's which resulted in a new model which worked well enough for Piazzi to suggest it as an item for the planned Exhibition of 1862. A proposal to ask the Government to fund the building of an oscillating platform at Portsmouth where the performance of the instrument could be properly tested was cancelled when Piazzi Smyth was disappointed with the work which Cooke's had done for him.

Piazzi Smyth's next move was to approach Sir William Armstrong, owner of the huge ordnance works at Newcastle. Armstrong was sufficiently interested in the revolver stand that he instructed that every assistance should be provided for Piazzi Smyth's job. The engineers, however, insisted that the design was flawed and that the fast-spinning wheel would inevitably tear away from its bearings. Piazzi Smyth tried to find a solution to the problem by putting the whole instrument on gimbals, but this would have made the project 'too expensive to be paid for privately' and he was forced to abandon the whole scheme on which he had worked so long and so lovingly.

The apparatus was praised for the ingenuity of its design. It pre-dated Thomson and Tait's demonstration gyroscope and appears to have been the first attempt to put the principle of the gyrocompass to practical use.[6] Copeland, Piazzi Smyth's successor in the Edinburgh post, who may well have seen the prototype in action, declared that 'it is much to be regretted that this important invention has never been fully developed'. More resources, financial as well as technical, would obviously have been needed to turn Piazzi's free revolver stand into a practical proposition.

In the Pulkova Observatory Piazzi Smyth had a particular interest in inspecting the main observatory clock which was kept in an underground room at constant temperature. He solicited the views of the Pulkova astronomers on the merits of 'gridiron' or mercurial pendulums to compensate for the effects of minor fluctuations in temperature on the lengths of pendulums and the corresponding rates of clocks. His interest arose, of course, from the time service in Edinburgh which he had started in 1852 with the operation of a time-ball on the top of the Nelson Monument on Calton Hill. Piazzi Smyth was not entirely satisfied with the efficiency of the time service until the appointment of a second assistant in 1855, since two people were needed to give the time signal, one to hoist the ball and the other to press the electric trigger at the clock at the correct instant. The use of electricity to transmit signals and the possibility of extending this over the telegraph system, was still a novelty. Greenwich had introduced it in 1852, transmitting signals simultaneously to an electric clock on The Strand in London and to the Admiralty dockyards.

At the Glasgow meeting of the British Association in 1855 Piazzi Smyth had demonstrated the advantages of electrical signalling by connecting a time switch at Edinburgh to a model time-ball placed in a hall in the University. The experiment, financed by Sir Thomas Makdougal Brisbane, was carried out with the cooperation of J P Nichol, the Professor of Astronomy at Glasgow University. Each day at the appointed hour Piazzi Smyth raised the ball; it duly dropped on the signal from Edinburgh and the time was compared with Nichol's observatory chronometer. No difference could be detected by the admiring audience.

The accuracy of the signal depended, however, on the clock, and Piazzi Smyth was delighted when Makdougal Brisbane offered to supply the Royal Observatory with a clock which was the best then available. On his way to the Paris Exhibition in 1855 Piazzi Smyth went to see Dent, the clockmaker in London from whom he acquired in 1856 a clock of the highest quality which became known as the Brisbane clock and which remained the observatory's main clock for decades. On the same visit to London Piazzi Smyth also conferred with Latimer Clark of the Electric Telegraph Company who was responsible for the connections between Greenwich and the outlying clocks in response to a suggestion by Airy that an attempt should be made to find the longitude difference between Greenwich and Edinburgh by exchanging time signals over the telegraph wires. The method had been developed in the United States for the American Coast Survey. The project was carried out very successfully in the course of one week in April 1858 through the simultaneous observation at Edinburgh and Greenwich of a set of 22 stars whose transits at the two observatories were recorded electrically in both places with the aid of an engineer from the Telegraph Company who cooperated 'con amore'. The time differences established the longitude of Edinburgh as 12 minutes 56.005 seconds west of Greenwich, correct to a hundredth part of a second.

In the years that followed Piazzi Smyth put a great deal of effort into improving the rate of the Brisbane clock; or, rather, into getting rid of external causes which sometimes gave rise to quite substantial variations in the clock rate. When he discussed the problem in Paris with Le Verrier he was shown an underground pendulum, carefully temperature controlled, with its clock face above ground. Piazzi Smyth considered placing his pendulum underground in the same style, but it seemed to him simpler to acquire one of the Pulkova 'gridiron' compensation pendulums; this, however, produced no significant improvement.

A real step forward came with the adoption in 1861 of the Jones system on the recommendation of John Hartnup, Director of the Liverpool Observatory and of Sir Thomas Maclear of the Cape. The system, whereby one clock could govern another electrically, had been invented by R I Jones of Chester and tried out at the Liverpool Observatory where a time service including a time-ball had been in operation for some time. Hartnup was a well-known astronomer who had for a while worked at Greenwich and had also been Assistant Secretary of the Royal Astronomical Society. Like Piazzi Smyth in Edinburgh, Hartnup had been aware of the need for accurate time in the busy port of Liverpool. He had investigated the effects of temperature on clocks and had set up an efficient chronometer rating service. He was the first to try Jones' suggestion of using two clocks for time signalling, a 'master' which actually controlled the time and a 'servant' or subordinate clock which was kept correct under the influence of the master and was the one which was connected electrically with other distant clocks. Thus the impulses to the distant clocks, or to the time-ball as the case might be, came not from the main clock but from the second one which did not have to be of the same standard as its master. As Piazzi Smyth clearly put it to his Board, 'the subsidiary clock is used as a servant merely to do the labour of striking the seconds; and the other, as its master, gives noiseless orders as to when they shall be struck'. Jones himself brought one of his clocks to Edinburgh to be used together with the Brisbane which could now be kept in a sheltered corner. The Jones arrangement was a precursor of the free pendulum clock which was constructed by W H Shortt and placed in the Royal Observatory, Edinburgh for trials in 1923. Shortt clocks became standard timekeepers in many observatories.

Meanwhile moves were afoot to extend the Edinburgh time service by the addition of a time-gun, an ambition which the Edinburgh Chamber of Commerce had cherished for years. The idea of having an audible as well as a visible time signal was not new: loud striking clocks were being installed elsewhere, the most famous being the clock on the Palace of Westminster ('Big Ben') which had been made by Dents under the supervision of E B Denison (later Baron Grimthorpe) and Sir George Airy. However, no other city in Britain had an audible signal as powerful as a gun. The Chamber of Commerce keenly subscribed to the gun project, the Government approved and offered a special gun and the professional help of the Army at Edinburgh Castle.

The ancient castle of Edinburgh stands on a dominant rock facing the Calton Hill across a deep valley at a distance of about a mile. The contract for the cable connecting the Nelson Monument on Calton Hill to the castle was given to the firm of Ritchie's, the Edinburgh clockmakers, who boldly decided to carry the wire in one single span with a massive steel cable weighing some three hundredweight. With a length of over 4000 feet, it was the longest open-air span ever attempted. To achieve this was a remarkable feat of engineering which was accomplished in only two days in April 1861. The actual work of hoisting the cable was performed by sailors from Leith who worked in the early hours of the morning, watched by Piazzi Smyth. The cable was at a minimum height of 240 feet above the ground and the scheme was possible only because of the peculiar geological

A view of Edinburgh Castle across the valley of Princes Street Gardens (Scottish Tourist Board).

Sailors on Nelson Monument fixing the wire connecting the observatory to the castle for the time-gun, drawing by CPS (RSE).

formation of the city where a steep valley separates the two hills. The gun was activated by a second Jones clock regulated by the Brisbane clock at the observatory, supplied by Ritchie's at their own expense and installed at the castle. Incidentally, this marked the first step towards international recognition of the Edinburgh firm of Ritchie's who soon afterwards supplied a pair of clocks to the Pulkova Observatory and later designed their own version of the Jones clock which was afterwards adopted in observatories all over the United States.

Experiments had to be made to find out how much gunpowder was needed to make the gun blast at the castle audible but not destructive. Piazzi Smyth and his assistants travelled all over Edinburgh and as far as Leith docks to test how well they could hear the sound. The gun was placed on the half-moon battery on the east side of the castle where it was highly visible from most parts of the city. The ships in the Forth found that whereas they required binoculars to observe the fall of the time-ball, they could see the smoke of the gun with the naked eye, and could have a second check on the time when they heard the actual signal a few seconds later. A time-gun map of the city was printed and distributed on which sound-second circles indicated the time lapse between the firing of the gun and the arrival of the sound of its blast.

The inauguration of the regular use of the Edinburgh time-gun was fixed for 5 June 1861, to coincide with the annual visitation of the Board of Visitors to the Observatory. Following their meeting the Visitors together with a large number of invited guests assembled outside the observatory to watch both the gun and the time-ball. Piazzi Smyth himself remained in the building attending to the clock. It soon became obvious that there was something amiss: the ball dropped but the gun was silent. The fault was with the firing fuse which was the responsibility of the Artillery at the castle, but this did nothing to lessen the acute embarrassment felt by the Astronomer Royal for Scotland in the presence of the dignitaries of the city and the university. 'The wine remained untouched', he noted pathetically in his journal that evening. The fault was, however, soon rectified; indeed the time-gun proved extremely efficient and the occasional non-firing happened on average only once in 70 times. Never once was there an error on the side of the observatory. The gun was a great success with both the public and the masters of the ships in port who declared it to be much more useful than the time-ball because it could be heard well outside the boundaries of the city.

The fame of the time-gun spread. In 1863 the Universal Private Telegraph Company approached Piazzi Smyth asking for time signals to be transmitted to Newcastle upon Tyne, a hundred miles away, for

First firing of the time-gun in 1861, drawing by CPS (EAO vol. 12). The dog in the picture is reputed to be the Skye terrier 'Greyfriars Bobby', famous for his devotion to his master's grave in Greyfriars Churchyard and immortalised by a statue which is one of Edinburgh's landmarks.

the purpose of firing a gun there. The engineer of the Telegraph Company, Nathaniel Holmes, was a man whose enthusiasm matched that of Piazzi Smyth, and by September of that year a gun on the Norman keep at Newcastle was successfully firing from the same signal which activated the gun on Edinburgh Castle. No less a person than Prince Albert came to Newcastle to witness the firing. Soon the local seaports on the Tyne—North Shields and Sunderland—wanted their own guns. Finally Glasgow and the nearby port of Greenock on the Clyde had similar guns installed, all receiving signals from the one source. Piazzi Smyth declared that provided clients brought their electrical wires to the Calton Hill, the observatory would provide the time, since it was no more trouble to give the time to all than to one.

The Director of the University Observatory in Glasgow at this time was Robert Grant, author of a scholarly *History of Physical Astronomy*, who had succeeded Nichol in 1859. Piazzi Smyth, assuming that like his predecessor he was an academic with little interest in technical matters, had not considered that the Edinburgh-fired gun would cause offence in Glasgow. Indeed he thought that as a government servant and Astronomer Royal for Scotland it was his duty to supply the time anywhere in the country where there was need for it. 'I quite innocently sent out from here the electric signals and gave that city the intellectual luxury and business accommodation of true daily time', he said. Grant saw things differently: he took it as an insult that Edinburgh should supply time to Glasgow and he set about providing time from his observatory to various clocks in the city, employing Ritchie's to provide the necessary clocks.

By 1864, not only the Glasgow gun but those in the Newcastle area had all fallen silent. The city authorities in Newcastle decided to give the contract for the electric signals to a different company who intended to get the time from Greenwich in the future. When nothing came of this, Piazzi Smyth concluded that the second company was not as competent to look after the technical part of the operation as the original company. In fact, the Electric and International Telegraph Company which already had links with Greenwich was planning to distribute Greenwich time to the railway network throughout Britain; this they achieved one year later, in 1865. With accurate clocks in every railway station there was no longer need for time-guns in normal civilian life. One gun only was left, the original one o'clock gun on Edinburgh Castle which is still fired daily after well over a century.

The Edinburgh time service was indirectly responsible for bringing into astronomy a man who was to become one of its leading lights and a staunch new friend of Piazzi Smyth. He was recommended to Piazzi Smyth by Professor David Thomson of Aberdeen who wrote in November 1863: 'I have been requested by Mr David Gill Jr whose father is a watchmaker in this place and who is desirous of introducing here the time-gun system, which owes its origin to yourself, to give him a few lines of introduction and to secure for him the advantages of your aid and advice in carrying his intentions into effect.' Piazzi Smyth was delighted to oblige the young man and so began a life-long friendship between him and the future Sir David Gill, Her Majesty's Astronomer at the Cape. David Gill was invited to Edinburgh to see the transit instrument and the time-gun, and soon he and Thomson succeeded in rigging up a transit instrument at King's College, Aberdeen, with which they observed transits of stars and regulated a clock which supplied time in the city. From this beginning David Gill took up astronomy as a serious hobby until, eight years later, he was invited by the young Lord Lindsay to take charge of the private observatory he was setting up at his estate in Dun Echt, not far from Aberdeen.

Without a suitable equatorial telescope at the observatory Piazzi Smyth used on occasion to visit amateur astronomer friends who possessed good telescopes of their own. One of these was E J Collingwood, a man of Piazzi Smyth's own age, who had a private observatory at Lilburn Tower in Northumberland with a six-inch refractor. Another was J W Grant, a keen astronomer who, on returning to his native Scotland after more than 40 years in India, had built a sumptuous observatory on his estate at Elchies on the River Spey in Morayshire. Here he erected a refractor of 11 inches aperture and 16 feet focal length, by Ross, on a substantial German mounting constructed by Ransome and May, the engineering firm. The telescope was at the time the largest in Scotland and only slightly smaller than the largest refractors in the British Isles which were the twelve-inch refractors at Greenwich and at Dublin. The next largest in Scotland was the nine-inch telescope at the Glasgow University Observatory. The Elchies telescope had occupied a conspicuous place at the Great Exhibition of 1851 in London; its maker Andrew Ross called it his 'trophy' when it earned him a medal at the Exhibition. The instrument in its ornate dome in beautiful scenic surroundings had lain idle for years until, in the autumn of 1862, Piazzi Smyth was invited to Elchies to have a look at it.

Piazzi Smyth's three-week sojourn at Elchies produced two substantial scientific papers and also inspired a sheaf of delightful paintings of Highland scenes. The most interesting aspect of the observations was the comparison which Piazzi Smyth was able to make between the practical performance of this relatively large and optically excellent instrument in a poor climatic location and that of the seven-inch refractor which he had used at Alta Vista on Tenerife. A number of double stars were observed, among them the well-known pair γ Andromedae which is made up of a bright yellow star and a less bright blue one separated by 10 arc seconds. In 1842 Otto Struve discovered that the blue component was itself double, the separation of this pair being less than one arc second. At Elchies there was only one occasion in a full three weeks when the second and third components of the system were separated whereas at Alta Vista they had been seen resolved every night without fail. This observation confirmed the great advantage of the mountain site for resolution and clarity, but Piazzi Smyth reiterated at the same time the need for a large-aperture telescope. A small instrument on a mountain could not do the work of a large one, as was sometimes claimed in a misunderstood interpretion of his Tenerife work. 'As yet', wrote Piazzi Smyth, 'such a station (i.e. Tenerife) has been worked with only a comparatively small size of telescope; and there seems to be a predeliction for confining it to such; under the hope, apparently, that the improved circumstances may more than counterbalance the want of intrinsic power. It would seem to be a wilful defeating of the very ends for which such an observatory should be founded, if a Mountain-stationed equatorial on a future occasion were not permitted to be fully equal in size and quality to the largest used at any of the lower observatories.' A large telescope and a good mountain site were both essential.

An unexpected observation made with the Elchies refractor was the apparent absence of one of the four stars in the system 222 Arietis (in the constellation of Aries, the Ram), a system which had been described in some detail in the *Cycle of Celestial Objects* and which Piazzi Smyth had observed from Alta Vista in 1856. At Elchies in 1862 the star, though one of the two brightest members of the group, could not be seen at all, a result confirmed a few months later at Lilburn Tower Observatory. The apparently mysterious change was due to the motions of the components in the system. Piazzi Smyth's dramatic account to the Royal Society of Edinburgh of the vanished star impressed itself on his listeners; a member of the 'Symposium', the professorial dining club, composed a song called 'The Lost Star', sung to the tune of Thomas Moore's 'I saw from the beach', which was rendered on more than one occasion to that convivial assembly:

> I saw from the Peak over Tenerife's seaside
> A star that in Aries most gloriously shone
> I looked through the great equatorial at Speyside
> The Ram was still there but the bright star was gone.
> Ne'er tell me of asteroids endless in series,
> Ne'er tell me of comets so blazing and hot.
> Give me back, give me back my bright sparkler in Aries,
> A twinkle from him's worth the whole of the lot.

At one of those dinners when the song was sung the club minutes record that 'the Symposists believing that nothwithstanding the extinction of this luminary by the Professor of Practical Astronomy, they would find as many stars as usual to light them on their way home tonight, adjourned after spending, according to their wont, a hilarious evening'.

The single observing run by Piazzi Smyth on the Elchies refractor represents the only astronomical work ever done with this excellent instrument, and illustrates the odd state of affairs whereby a professional national observatory had no useful telescope for extra-meridian work, while wealthy men throughout the country could afford to set up private observatories, many—as at Elchies, Lilburn Tower, Hartwell and the Smyth family's former observatory at Bedford—possessing telescopes, transit instruments, clocks and all the accoutrements of professional astronomy. The fashion for private observatories owed a great deal to the immense popularity of Sir John Herschel and the 'contagion of his example'. While some of these private observatories undoubtedly contributed to the fast-increasing store of astronomical information in the nineteenth century there were others which were no more than adornments to wealthy gentlemen's country houses. It was indeed strange that Piazzi Smyth could only look through a good telescope through the hospitality of his amateur scientist friends.

An equatorial telescope was not the observatory's only deficiency. The whole establishment was by the 1860s in dire need of renewal. When Piazzi Smyth negotiated the transfer of the observatory to the Government in 1847 the necessary repairs to the building were carried out and the two meridian instruments were put back into good working order, but the idea of an annual budget and regular funds to maintain the instruments had not entered into the discussions. The Government had made only occasional grants to pay for printing the *Edinburgh Observations*. Now, after years of constant use the instruments were no longer up to standard. For new equipment the observatory had to rely on gifts, such as the badly needed clock received from Sir Thomas Makdougal Brisbane and a portable chronometer from Sir William Gibson-Craig. The expenses of Piazzi Smyth's scientific excursions were met in a similar fashion, being funded either out of his own pocket or through the kindness of friends. It is an extraordinary fact that it was only in 1861, fifteen years after Piazzi Smyth had taken office, that an official request was made by the Treasury for a statement of the observatory accounts. This revealed that Piazzi Smyth was owed more than £100 by the Treasury!

There was also a serious problem of accommodation. Playfair's 'Temple of Urania' was one of the ornaments of the city, regarded by the people of Edinburgh 'from a picturesque rather than from an astronomical point of view'. Piazzi Smyth could point out to his Board that the smallness of the premises meant that there was only one computing room which had to accommodate two people as well as the much swollen bulk of observatory papers and the publications arriving in exchange from other observatories. It had also to serve as a library though it was entirely unsuitable for books; it was damp and dirty on account of the open fire and the gaslights which burned at night. There was no room of any sort to store equipment. The solution proposed by Piazzi Smyth was that the government should provide a larger official residence for himself, nearer to the observatory than the existing one, which could double as office and storage accommodation. There was no reaction to this proposal from the government.

In 1862 a new member appeared on the Board of Visitors who became Piazzi Smyth's staunchest supporter. This was Peter Guthrie Tait who in 1860 succeeded Forbes in the Chair of Natural Philosophy at the University of Edinburgh. Forbes, who had been in poor health and had rarely appeared at Board meetings for some years succeeded Brewster as Principal of St Andrew's University, Brewster having been appointed Principal of the University of Edinburgh. The eminent Brewster thus became an *ex officio* member of the Board of Visitors. Tait was from the first an active and conscientious member of the Board—as he was in his role as Secretary of the Royal Society of Edinburgh in later years—and became its Secretary. He was one of the brilliant new generation of physical scientists, schoolmate and lifelong friend and correspondent of James Clerk Maxwell, and collaborator with Thomson (Lord Kelvin), his opposite number in the Glasgow Chair, a collaboration which led to the famous fundamental treatise on thermodynamics, *Treatise on Natural Philosophy*, usually referred to as 'Thomson and Tait' published in 1867. Tait was, furthermore, a man of great personal charm and good humour; it is admirable to see how well he could handle Piazzi Smyth's often idiosyncratic methods and smooth out his wilder outbursts in later years.

The meeting of the Board of Visitors of October 1863 was a particularly significant one in the history of the observatory, as it was on that occasion that the role of the observatory and the support necessary to sustain it were clearly spelled out for the first time. It was Tait's first meeting as Secretary, and Brewster was also present, in one of his rare appearances. At the instigation of R M Smith, representing the Edinburgh Chamber of Commerce, a resolution was passed calling on Her Majesty's Government to implement the promise made in 1848 'to maintain the Observatory of Edinburgh in all time coming as a "proper Royal Observatory"'. An increase in

Peter Guthrie Tait. Drawing by William Hole, ARSA (Quasi Cursores, Edinburgh 1884).

current expenses, a grant of at least £400 for repairs to instruments and for publication of observations, and an increase in Piazzi Smyth's salary were asked for—'the post of Astronomer Royal for Scotland, entailing far more continuous and laborious exertion than any Professorship, is absolutely without salary; the Board consider that the emoluments of the conjoined offices are much too small'.

Duncan Maclaren, Chairman of the Edinburgh Chamber of Commerce, a former Provost of the city of Edinburgh and later an energetic Member of Parliament for Edinburgh, took it on himself to go to London and make representations personally to the Lords of the Treasury on behalf of the observatory. An official examination of the building and instruments took place and an estimate for repairs was ordered. At last it appeared as if something might be done; but the years slipped past and the observatory was still ignored by the authorities in London.

Chapter 6

The Metrology of the Great Pyramid

In 1863 Piazzi Smyth, by then in his forties, had been 17 years in office. In that time he had not only completed the reduction and publication of his predecessor's observations, but had also initiated a major new programme in positional astronomy. He had inaugurated an extensive time service and had introduced a university teaching course. Most important of all, he had carried out his successful site-testing expedition to the Canary Islands which was acclaimed in the highest international quarters. However, in spite of his efforts to arouse a practical interest in mountain astronomy at home, and in spite of his Board of Visitors' unfailing support, the outlook for the observatory was bleak. It is not surprising that Piazzi Smyth, still full of vigour and enthusiasm, was looking around for a new challenge.

It was unfortunate that the task which he then took up was a decidedly bizarre one which arose out of his recently acquired interest in the metrology of the Great Pyramid of Gizeh (or Giza). The story of ancient Egypt had in Piazzi Smyth's days become a subject of much curiosity following the exploration of Egyptian antiquities by 175 French 'savants' who had been ordered to accompany Napoleon in his campaign of 1798. They had published their various discoveries in a set of more than 20 volumes of text and plates which created great enthusiasm amongst archaeologists everywhere. The pyramid had been visited by Napoleon himself in July 1798, a month after his victory over the Mamelukes in the Battle of the Pyramids. Like Alexander the Great he had insisted on being left alone in the mysterious surroundings of the King's Chamber. The pyramid, the only remaining building of the Greeks' 'seven wonders of the world' had frequently been thought of as enshrining profound secrets about man and his destiny. It had also been looked at as a structure which embodied fundamental laws of geometry which later passed from ancient Egypt to the Greeks. With such ideas in mind one of Napoleon's savants, E F Jomard, had studied both the external dimensions of the pyramid and the geometry of its interior passages and chambers. He had suggested that an empty stone chest which had been found in the King's Chamber was not a sarcophagus of an ancient Pharaoh, but some measure of capacity.

The pyramid has a simple shape with a square base of side about 230 metres, and a height of about 145 metres. Its sloping faces are inclined at an angle of 52 degrees to the horizontal plane, a fact which leads to the existence of certain striking relations between its various linear dimensions. More than one of these had been noticed by observers in earlier days, but the investigation of their exact nature and ultimate origin became the centre of Piazzi Smyth's ideas and of the counter arguments of his opponents. The dimensions of the pyramid are such that the ratio of the length of one of its sides to its height is close to 11/7 or to half of the ratio 22/7 which since the days of Archimedes in the second century BC has been recognised by mathematicians as a first approximation to the ratio of the circumference of a circle to its diameter. Another fact about the pyramid had already been known to Herodotus in the fourth century BC who recorded in his *Histories* that the area of each of the pyramid's faces is equal to the square of its height.

The Great Pyramid of Gizeh, photograph by CPS (RSE).

Piazzi Smyth was particularly intrigued by the first of these relations, that the perimeter of the base of the pyramid appeared to bear the same relation to its height as the circumference of a circle to its radius. To him this indicated that the builders of the pyramid had been familiar with the numerical value of the number pi (π) and that they had in fact designed the pyramid so as to demonstrate to future generations that they had solved the age-old mathematical problem of 'the squaring of the circle'. Since it was well known that the pyramid had been built long before the Greek mathematicians or their Egyptian predecessors had grappled with that problem, Piazzi Smyth became convinced that the pyramid in its particular form could not have come into existence by unaided human efforts.

Such a suggestion came oddly from a man like Piazzi Smyth who was no novice to Egyptology and who looked with admiration at his father's efforts in this field. W H Smyth had become a knowledgeable archaeologist during the years which he had spent in the Mediterranean. His particular interest in ancient Egypt had led him to early discussions with his friend Mehmet Ali, 'the Pharaoh of the 19th century', on the possible move from Egypt to England of 'Cleopatra's Needle', the obelisk which had been taken by Augustus to adorn the Caesareum at Alexandria and which in fact was removed from there to London in 1877. Smyth's enthusiasm for ancient Egypt led him to name one of his daughters Rosetta after the celebrated stone found near the mouth of the Nile which had supplied Champollion with the key to the deciphering of the Egyptian hieroglyphs. In his *Aedes Hartwellianae* Smyth devoted one entire chapter to Egyptian antiquities in which he roundly ridiculed the speculations of 'the sages who perceived an indication of the mariner's compass in the mystic Tau, the symbol of eternal life, and of a still more inducted set who believed that the Pyramids were erected for the squaring of the circle'. Piazzi Smyth himself was well familiar with the work of Egyptologists such as C R Lepsius who had led a Prussian archaeological expedition to Egypt in the 1840s and with the work on Egypt of C C J Bunsen and others. His interest in ancient Egypt showed clearly in the lectures which he gave to students in the University of Edinburgh in the 1850s.

The change in Piazzi Smyth's thinking from sober archaeology to mystical speculation was the result of an odd combination of circumstances. It originated from the writings of a certain John Taylor who had become obsessed by the structure of the pyramid and who at the same time held strong views on what he considered proper standards in metrology. His pamphlets were coloured by a religious fervour for the propagation of the truths of the Old Testament. Piazzi Smyth himself being a devout Christian with evangelical leanings was of one mind with his wife who was deeply struck by Taylor's 'virtues and Christian piety'.

Taylor was a publisher,[1] a partner in the London firm of Taylor & Hessey, who had also acted as editor of the *London Magazine* which published contributions from writers such as Lamb, Coleridge and Keats. Taylor's own writings had originally been mainly political and with his views on standards in fiscal policy he had taken up his pen in opposition to Sir Robert Peel's Currency Bill of 1844. In his later years he became attracted by Colonel Howard Vyse's account of his exploration of the pyramid, published in 1837. Using Howard Vyse's measurements Taylor started his own calculations and convinced himself that the π ratio between the circumference of the base of the pyramid and its height was correct. He also confirmed the Herodotus statement that the area of each face of the pyramid equalled the square of the height. He was profoundly impressed by these numerical findings which, he felt, could not be due to chance. Agreeing with Jomard's pronouncements Taylor also took the view that the builders of the pyramid had possessed the necessary astronomical knowledge to measure the length of a geographical degree and thereby the circumference of the Earth. As Taylor put it, 'it was to make a record of the measure of the Earth that the Pyramid was built'.

Taylor further suggested that in the actual construction of the pyramid the builders had used cubits of close to 25 British inches long as units of length, and concluded therefore that the British inch represents a measure of length which is of very ancient origin. In this view Taylor was following some distinguished predecessors, chief among them Isaac Newton who had studied the metrological work of John Greaves, Savilian Professor of Astronomy at Oxford, who in the early seventeenth century had made measurements of buildings in Greece, Rome and Egypt. Newton had also taken into account Flavius Josephus' statements on the dimensions of the Temple of Jerusalem and had published his conclusions on ancient standards of measurement in a Latin dissertation which was translated into English by Thomas Birch in 1737 and was reproduced by Piazzi Smyth in his own writings. Newton suggested that the 'vulgar cubit of Memphis' which had been quoted by John Greaves was about five inches shorter than the 'sacred cubit' which according to the Old Testament had been used by Noah in the building of the Ark and by Moses in the construction of the Tabernacle. It was this sacred cubit, 25 inches long, which, according to Taylor, had also served the builders of the pyramid as a standard of length. He also claimed that the pyramid contained an independent standard of capacity in the form mentioned already by Jomard of a vessel or 'coffer' in the King's Chamber in the pyramid. As far as Taylor was concerned, the pyramid, which he believed to have been constructed during the Israelites' sojourn in Egypt, was a building whose principal purpose was the preservation for all time of the sacred standards of length and capacity which had been delivered to mankind in the writings of the Old Testament.

These strange ideas were submitted by Taylor to the Royal Society of London in a paper entitled 'A memoir of the Great Pyramid. Why was it built?' Not surprisingly, the paper was rejected by the Society, but Taylor published it in amplified form in a book called *The Great Pyramid; Why Was It Built? & Who Built It?* He distributed copies of this book amongst various influential people who, he thought, might acknowledge his ideas, among them John Herschel. Piazzi Smyth received a copy through a common friend, Patrick Scott, a poet and writer on religious subjects. Piazzi Smyth's first response to the book which reached him in November 1859 was polite but distinctly unenthusiastic. In view of the change which took place later in Piazzi Smyth's views it is interesting to quote part of the letter which he wrote to Scott on 16 November 1859:

> There is so much study and learning of the subject displayed by Mr Taylor that it would be presumptuous for anyone who had not also gone deeply into the matter to sit in final judgement upon him. I will therefore only venture the general remarks or questions, why are the two greatest authorities on Egypt, Lepsius and Bunsen, not alluded to? Is it because Bunsen's remarks place the date of the building of the Pyramid several thousand years earlier than the Noachian deluge? That goes against Mr Taylor's ideas, but is apparently founded on an unbiassed consideration of all the monuments of Egypt; and to consider one of the monuments by itself does not seem to me the right mode of proceeding in the present state of Egyptology.

The letter closed with Piazzi Smyth informing Scott that he would pass Taylor's book on to the Architectural Institute of Scotland, and there the matter seemed to end.

Piazzi Smyth's original reaction to Taylor's views was therefore entirely consistent with the opinions which his father had expressed when he talked about 'erroneous conclusions stamped with factitious erudition'. However, when five years later Piazzi Smyth submitted his annual report for 1864 to his Board of Visitors his views had changed completely. He now claimed that the further he proceeded in his enquiry into Taylor's theory 'the more some of Mr Taylor's leading views seemed to commend themselves'. Later in the same report he states that 'a high probability was impressed upon my mind that the Great Pyramid, besides its tombic use, might have been originally invented and designed to be appropriate for no less than a primitive Metrological Monument'.

The immediate reason for Piazzi Smyth's conversion to Taylor's ideas was his reading of a second pamphlet of John Taylor's called 'The battle of the standards', published in 1864, which dealt with the problem of the use of proper weights and measures, claiming the superiority of the ancient British standards and their link with the sacred system of the Bible.

On the question of standards Piazzi Smyth himself was at all times in favour of the preservation of the Imperial and against the adoption of the French metric system which was widely advocated in public debates in the middle of the nineteenth century. This is clear from the discussion he had with the like-minded Otto Struve in Pulkova in 1859. In this matter of standards Piazzi Smyth staunchly followed his idol John Herschel whose opposition to the metric system was so deep that he resigned from membership of the government's Standards Commission when it came down in favour of the metric system. In 1863 a Bill favouring the introduction of the metric system had been put before Parliament. Before the debate on the Bill Herschel distributed to all members of both Houses of Parliament a pamphlet on the 'British modular standard of length' in which he argued in favour of the retention of the British system on scientific grounds. The Bill was not passed on that occasion but came up again in the following year as a permissive measure. Herschel continued to warn against the metric system on all possible occasions. In his widely read *Popular Lectures* he declared that

> it appeared to the Author an imperative duty to let no opportunity pass of calling the attention of the public to the great question of the proposed abandonment of our national system of weights and measures and adoption in its stead of the metrical system of the French with its unit, the metre, in place of the English yard, which has been so actively and in his opinion so mischieviously, urged on Parliament; the agitation in favour of which only sleeps for the present in the view of allowing the public mind to familiarise itself with the idea under a 'Permissive Act',—to be assuredly brought forward again with renewed activity and under a more intense and prolonged pressure on no distant occasion.

John Herschel objected to the metric system on the grounds that the formal definition of the length of the metre—the ten millionth part of a quadrant of the Earth—was far from satisfactory because it depended on the length of a particular meridian, originally that running through Paris, of an Earth which was not strictly spherical. Herschel asserted that the straight polar axis of the Earth would provide a better standard of length than a curved meridian, and that such a standard would have a unique length, 'the same for any nation on the globe'.

Using the results of the geodetic surveys available to him—which, of course, were ultimately based on measurements of base-lines made on the curved surface of the Earth—Herschel arrived at a figure of slightly more than 500 million British inches for the length of the polar axis. According to him, a value of 500 500 000 inches would in fact represent the length of that axis to better than one part in 60 000. Herschel pointed out that if the length of the normal British inch were to be increased by a thousandth part, or the thickness of a human hair, the Earth's polar axis would turn out to be exactly 500 million British 'geometrical inches' long and the new 'geometrical inch' would become a truly scientific, Earth-commensurable unit of length. The required minute change in length would obviously have no effect whatever on measurements in ordinary life. As a larger measure he proposed a 'geometrical cubit' made up of 25 such inches.

In his pamphlet on the 'British modular standard of length' Herschel, unfortunately, acknowledged his indebtedness to Taylor for drawing his attention to the commensurability of the length of the Earth's polar axis and that of the British inch, though he was careful not to fall in with Taylor's general ideas about the pyramid. In fact, commenting on Taylor's assertions about the value of π revealed in the pyramid, he wrote: 'The coincidence is certainly very striking. It by no means follows, however, that the ancient Egyptians were in possession of any *calculus* by which they could have arrived at a theoretical knowledge of the true ratio. It would be easy for people in possession of such appliances as they must have had at command, to ascertain the ratio in question by tracing on a flat pavement a circle of 100 feet in diameter, and actually measuring the circumference.' Piazzi Smyth saw no contradiction between Herschel's comments and the view he had come to take that the pyramid was a 'π-memorialising' structure and the earliest demonstration known of the 'squaring of the circle'.

The ancient problem of the 'squaring of the circle' had held an irresistible attraction to men with a certain kind of mentality. As E W Hobson[2] once pointed out: 'The man of mystical tendencies has been attracted to the problem by a vague idea that its solution would, in some dimly discerned manner, prove a key to a knowledge of the inner connections of things far beyond those with which the problem is immediately concerned.' John Taylor, a man with little or no knowledge of mathematics, was a good representative of the type of person which Hobson had in mind, but it was sad that Piazzi Smyth succumbed to similar confusions and that he mistook Herschel's acknowledgment to Taylor as conferring somehow scientific status on the pyramid theory. It must also be said that Herschel's insistence on the commensurability of the inch with the diameter of the Earth was not a happy one. There is little doubt that Piazzi Smyth's adulation of Herschel had a lot to do with his unfortunate incursion into Taylor's scientifico-religious world.

Piazzi Smyth never met John Taylor who died in 1864, but correspondence between them shows that Taylor was overjoyed that an important astronomer had taken up his cause and that his labours had therefore not been wasted. In the last months of Taylor's life Piazzi Smyth was already busy writing his *Our Inheritance in the Great Pyramid*, a large tome whose first edition appeared in that same year. There were to be four more editions of which the last one was published as late as 1891. This immensely popular book was dedicated 'to the continued memory of the late John Taylor' by 'the friend of his last few days, but admirer of all his long and earnest Christian life'. It contains a detailed account of Piazzi Smyth's own version of Taylor's writings which otherwise would have disappeared into oblivion. Piazzi Smyth was always at pains to present Taylor as the originator of the religious aspects of pyramid theories, endowing him with the aura of a prophet, though, in fact, mystical ideas concerning the pyramid were rife and Taylor was by no means the only one to promulgate them.

It was Piazzi Smyth's views on the fundamental importance of metrology and on what he conceived as his duty to uphold proper metrological standards which led to his decision to go out to Egypt for a new exploration of the pyramid. His aim was to re-measure with the greatest possible precision the pyramid's external size and the dimensions and general geometry of its interior passages and chambers. In this particular task he was to succeed splendidly. Piazzi Smyth was also interested in testing the evidence for an earlier suggestion of John Herschel that it might be possible to date the building of the pyramid if one could assume that the direction of the sloping entrance passage was of astronomical significance. The French explorers of Napoleonic days had already found that the four sides of the pyramid had been aligned with considerable precision in the north–south and east–west directions. The entrance passage in its northern face appeared to point to the celestial pole, the centre of the daily rotation of the heavens. This pole being a projection on the sky of the terrestrial pole follows the slow conical, so-called precessional, motion of the Earth's axis which makes it move amongst the stars with a period of 26 000 years.[3] At present the star α Ursae Minoris, the Pole star, is close to the celestial pole, but on account of precession that same star in ancient times was a considerable angular distance away from the pole. Herschel, at the request of Howard Vyse, had computed that the celestial pole would have been marked by another reasonably bright star, α Draconis, at around 2160BC. Assuming then that the sloping entrance passage to the pyramid was intended to point to the celestial pole and that the star α Draconis was the pole star at the time, the pyramid must have been built at about 2160BC, making it younger than the Egyptologists of Piazzi Smyth's time believed.[3]

Our Inheritance in the Great Pyramid, Piazzi Smyth's popular book, was in its own way a masterpiece. Written and published within only half a year, it was an immediate success. Unlike Taylor's dull and confused volume, *Our Inheritance* carried its readers along as through an exciting adventure. It opened dramatically by naming the 'Three Keys required for opening the Great Pyramid': pure mathematics (the value of π), applied mathematics (astronomy and metrology) and human history (the Bible). The historical and archaeological background was well researched and documented lending weight to the book's 'scientific' parts in the eyes of the many religious-minded but non-scientific readers. There were hints—such as the quotation of the number π to 75 decimal places and numerous abstruse-looking diagrams—that profound mathematics were involved. Not that Piazzi Smyth had any intention of deceiving his readers; he had become utterly convinced that the theory had biblical authority and his intention was to demonstrate that the facts supported the theory. Piazzi Smyth's wife later wrote that after completing the book 'he felt that, if possible, he must eventually go to Egypt and re-measure the grand old Monument, whether his observations should affirm, or whether they should refute, the wonderful theory of the Great Pyramid'. However, from his statements in the book it is clear that his mind was already made up.

The geometrical and numerical part of the same material was described in a 40-page paper entitled 'On the reputed metrological system of the Great Pyramid' which Piazzi Smyth read to the Royal Society of Edinburgh in March 1864. Like all his papers it was meticulously prepared. All existing records of the pyramid were thoroughly researched and discussed, details of the apparently significant geometry of the inner passages were described and there was a section on chronology. As a piece of scientific investigation, however, it was seriously flawed. The outer dimensions of the pyramid were shown to be consistent with the π hypothesis, and indeed this was the case to within an accuracy of a few per cent. However, the analysis went much further. By suitable weighting of the data from various sources a far too exact agreement, entirely unjustified by the data, between the ratio incorporated in the pyramid and the number π was derived. Again there was no deception on Piazzi Smyth's part: he made no secret of his line of argument which was that where observations differed from theory, the observations were at fault. 'The Pyramid has proved itself to have been built closer to the truth than the best and greatest of modern savants from all existing civilised countries have been able to measure it: and if we are obliged to draw rein in pushing on the severity of our inquiry to still further places of decimals, the reason is not that the Pyramid fails in accuracy, but that modern observations are not sufficiently good to bear more than has been already put upon them.' The Fellows of the Society who listened to this eloquent presentation, perhaps mesmerised by the numerous tables and diagrams, made no adverse comments at the time, and the paper was published unchanged in the *Transactions* of the Society.

Piazzi Smyth announced to the Board of Visitors at the next meeting that he proposed to go to Egypt at his own expense to make measurements at the pyramid. The reason he put to the Board was concerned mainly with metrology, 'seeing how our nation is agitated just now by questions of a change in its hereditary metrology', and the need to investigate how the hereditary measures acquired 'a primal resemblance to those of the Great Pyramid and to some of the sacred Hebrew measures of old'. The minutes of the meeting do not record the reaction of the Board. The only official support he had was a letter from the Secretary of State for Foreign Affairs, Lord Russell, to the Consul-General in Egypt requesting an introduction for Piazzi Smyth to the Viceroy.

Piazzi Smyth prepared very carefully for his scientific mission to the pyramid. He designed and ordered from Adie's of Edinburgh and Cooke's of York a set of measuring rods of various sizes from tiny finely graduated ivory scales readable to a hundredth of an inch up to long wooden rods. For measurements of interior dimensions he obtained from Cooke's a number of sliding rods made of mahogany, with steel tipped brass pointed ends, which 'gave great satisfaction' in actual use. A friend in Manchester, Joseph Sidebotham, a factory owner who was also an amateur photographer, astronomer and naturalist, donated a rod made of an organ pipe of Queen Anne's time, and therefore well seasoned, encased in its own box. Unfortunately the instrument-maker who had been entrusted with the task of making the rod had oiled it, so that it warped under the Egyptian sun! However, Piazzi Smyth substituted an even better measure in the form of an ancient piece of basalt which he marked at each end by fine scratches made with Jessica's diamond ring. This stone measure was calibrated on their return from Egypt against one of Kater's standard yard sticks.

For astronomical and angle measurements Piazzi Smyth had three good instruments: Playfair's splendid altazimuth telescope, a sextant designed by Piazzi Smyth and constructed by Adie's for the Paris

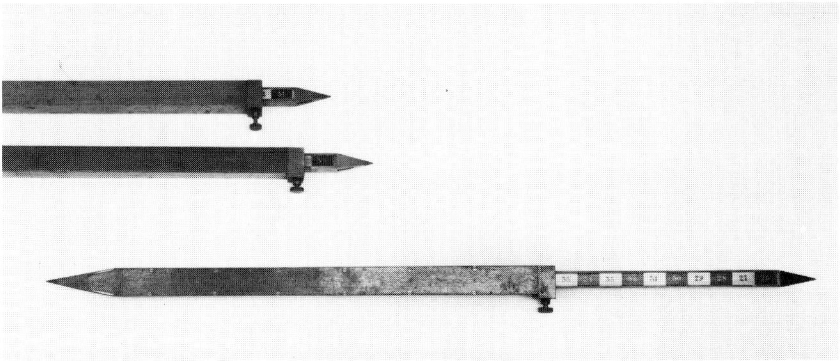

Piazzi Smyth's adjustable pyramid measuring rods (RMS).

Playfair's altazimuth telescope, used for astronomical observations at the pyramid (RMS).

Exhibition of 1855 which employed a spirit level instead of an artificial horizon for levelling; and a clinometer with an eight-inch circle, again designed by himself and constructed by Cooke's especially for the pyramid work, which was donated by a friend.

There was also an assembly of meteorological instruments, all tested by Hartnup at the Liverpool Observatory. Piazzi Smyth planned to record as much as possible photographically, and he designed and constructed with the help of a local carpenter a novel box camera to take small plates to be processed *in situ*. He took along also a conventional camera and a stereoscopic one, and magnesium ribbon to use for flash photography in the dark interior of the pyramid. Thirty boxes were needed to carry all the necessities of the expedition,

Piazzi Smyth's pyramid camera and miniature plateholder (ROE).

which included a tent, domestic equipment and a supply of candles and books, in addition to the scientific instruments.

The Smyths sailed from Holyhead at the beginning of November 1864 by passenger steamer for the three-week voyage to Alexandria, in very ugly weather. The ship rolled and pitched ceaselessy, but while most of the passengers stayed below in their cabins, Piazzi Smyth remained alert, watching the unruffled crew going about their business and noting every unusual occurrence in his notebook. He made drawings of sea and sky, of fish that were washed onboard, and of a near shipwreck of a passing vessel which had lost its mast. In the calmer waters of the Mediterranean he painted many lovely watercolours and recorded in words the changing hues of the light on the water and in the sky.

He also found time to do some astronomical calculations, to read books on Egypt and to study religious tracts which more and more convinced him of his own divinely guided mission to discover the mysteries of the pyramid. In one place he has a comment: 'Ephesians 3. 9, 10, 11 applicable to Gt. Pd. being purposely kept concealed or misunderstood until a certain time before appointed.' The reference is to St Paul's words: 'To me who am less than the least of all God's people He has granted of his grace the privilege of proclaiming to the Gentiles the good news of the unfathomable riches of Christ and of bringing to light how the hidden purpose was to be put into effect. It was hidden for long ages in God the Creator of the Universe in order that now through the Church the wisdom of God in all its varied forms might be made known to the rulers and authorities in the realm of heaven. . . .'

The ship stopped for a day at Valletta in Malta, and the Smyths took the opportunity of calling on the astronomer William Lassell at his observatory a short distance outside the town. Lassell, a man about 20 years older than Piazzi Smyth, was one of the last of the great amateurs in the tradition of William Herschel, a self-made man who constructed his own instruments with such skill and used them to such purpose that he was accepted in the highest professional circles and honoured with a medal from the Royal Astronomical Society and a doctorate from Cambridge University. He rivalled Lord Rosse in the construction of reflecting telescopes and was the first to mount a large reflector on an equatorial mounting. He had a steam-operated machine made for him by James Nasmyth to grind his mirrors and his greatest achievement was a four-foot mirror which he used in Malta to observe faint nebulae. This massive telescope used an open-work tube in the modern way instead of a solid cylinder and was driven equatorially by a hand-driven mechanism synchronised with a seconds pendulum. Piazzi Smyth was shown the impressive observatory with its huge workshop, the polishing machine, lathes, benches, innumerable tools, supplies of wood and metal bars. The telescope with its open tube, Lassell explained to Piazzi Smyth, performed decidedly better than one with a solid tube by eliminating most of the 'twirling and twitching' of a star's image. An observing platform with three separate stories was capable of being moved around on a circular railway.

The Lassells' house was as lavish as the observatory and the Smyths were hospitably entertained by Mrs Lassell and her daughters. However, when Jessica introduced the subject of the pyramid, their host made it quite plain that he considered their mission a waste of time. 'So we left them', noted Piazzi, 'glad to have seen them and obliged to them, but with a something, somehow lacking in mental satisfaction.' Poor Piazzi! To his mind the pyramid was an entirely sensible scientific phenomenon; once converted to it himself, he could not understand how anyone could imagine otherwise.

In Cairo, Piazzi Smyth was certainly not regarded as a crank. He was the first surveyor to come to the pyramid since Howard Vyse 30 years earlier and he received every cooperation from the British Consul and from the Egyptian authorities. He had the honour of an interview with the Viceroy, the powerful Ishmail Pasha, at his sumptuous residence, who assured him of his permission to survey at Gizeh, leaving details to the archaeologist Marietta Bey, the director of the museum at Boolak. Marietta was a Frenchman who had been in Egypt for many years and had been honoured with the title of Bey. Piazzi Smyth approved of Marietta's strict attitude regarding the conservation of archaeological remains; after past plunderings there were firm controls on the activities of investigators in Egypt.

The Smyths had to remain for nearly a month in Cairo while arrangements were being made on their behalf. It was the time of the American Civil War and there was 'cotton fever' in Egypt, with money pouring in to buy cotton. Piazzi Smyth had not realised the drastic effect this would have on their finances; it was, in fact, inflation on a major scale and the Smyths found their funds running away while they waited impatiently to get started. Jessica succumbed to fever, and in their sickroom in the hotel they were plagued with flies by day and mosquitos by night. At last the fearless pair were on their way on the backs of donkeys, with their supplies of charcoal, crates of fowl, candles and kitchen equipment. Marietta had arranged the transport for them; tents were to be supplied and the local villagers had been instructed to provide food at local prices. 'It called forth some courage in a lady to set forth on such a journey, but the spirit of her who was here concerned rose equal to the occasion, and encouraged itself by indulging the firm conviction that some turn in our fortunes would surely take place.'

On that first day their fare was buns washed down with muddy Nile water, but they had no complaints. With characteristic optimism Piazzi remarked: 'Did not that glorious old Roman, Pescennius Niger, silence his murmuring legions by exclaiming "What! Crave you wine when you have Nilus to drink of?"' One must admire the couple, no longer as young as they were on the Tenerife expedition, throwing themselves with such enthusiasm into their new adventure.

An adventure indeed it was which they both thoroughly enjoyed. They were lucky with the Arab helpers who offered their services at the pyramid—Alee Dobree, who had served Howard Vyse 30 years earlier and who knew every stone in the pyramid, and Ibraheem who with a bevy of assistants took charge of their housekeeping. These men persuaded them to forget the tents and to set up home, as Vyse had done, in one of the tombs. They chose one which had once been

'The Lady of East Tombs at home', Piazzi Smyth's title for this photograph of his wife (RSE).

used by Marietta Bey. 'Pray look here at this nice little tomb just opposite your door' said Ibraheem, 'I'll make that into my kitchen, and we'll have those two mummy pits that lie between filled up and then I can put my water filters there in a quiet corner shaded from the sun; and if you go down three steps, what another beautiful room you come to, it will just do for the lady and gentleman to take their dinner in.' So the Smyths settled into their suite of tombs and were excellently looked after and fed for four months.

Piazzi Smyth assisted by Jessica and Alee Dobree worked hard day

The Smyths' cook at their pyramid home, photograph by CPS (RSE).

and night. They made thousands of measurements in all. On the outside of the pyramid the sides had to be measured, then the height from base to summit which involved measuring and individually recording the height of each of the 200 steps, a task which took three full days. Interior measurements were even more difficult; the Smyths and their little crew had first to clean out the inner chambers where work had to be done by candle-light and with constant interruption from tourists. They measured not only the length, breadth and height of each interior apartment, but the diagonals and all possible dimensions of individual blocks of stone. The measurement of the King's Chamber alone occupied five days. In the evening Piazzi Smyth wrote up his log books in his 'stone bower'.

Astronomical observations of the Sun were needed daily for timekeeping, and at night the positions of stars were observed for the purpose of getting azimuths and latitude. Angular measurements were made of the elevation of the summit, and of the slopes of the various passages, effected by viewing a source of light—a candle shining through a small hole—at the other end.

Another branch of observation, faithfully followed though not of particular scientific value, were meteorological readings of thermometer, barometer and hygrometer—recorded inside and outside the edifice in order to be able to compensate for possible changes caused by these effects. They also tried to identify the types of stone found in different parts of the interior—this was Jessica's special field—and collected samples of these and of fossil shells which were later presented to the Royal Scottish Museum in Edinburgh. Piazzi Smyth also took a large number of photographs, showing scenes of the pyramid and the surrounding views which were of great general interest. His interior photographs were in fact the very first ever taken of the interior of the pyramid and the first ever successful flash photographs obtained outside a photographers' studio.

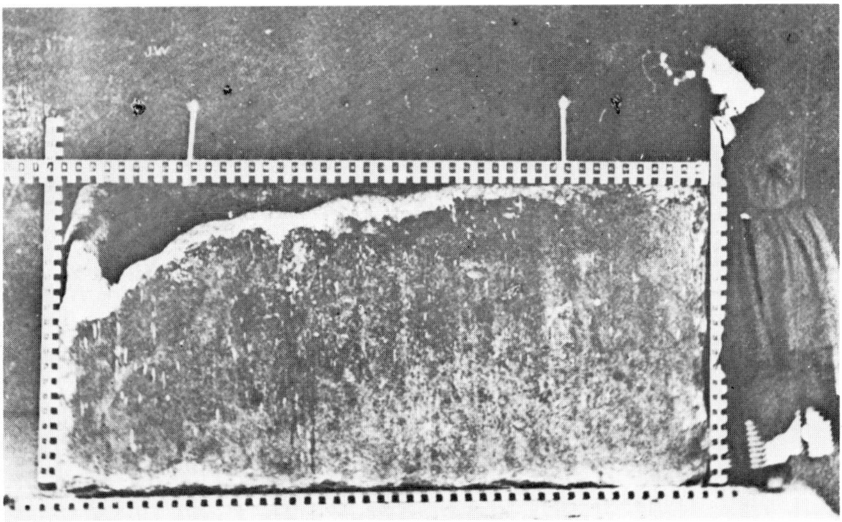

Interior flash photograph by Piazzi Smyth of the 'Coffer' in the King's Chamber being measured. Mrs Jessica Smyth is standing at the right (RSE).

One of the most important aims of Piazzi Smyth's expedition was to measure the exact length of a side of the pyramid, with a view to establishing the unit of length used in its construction. He realised that there was the problem of deciding at which vertical level such measurements should be taken. It was on record that Napoleon's French savants had uncovered cornerstones or sockets which would have marked the limits of the pyramid's base. Howard Vyse had in his day uncovered some of these but by the time that Piazzi Smyth arrived they had once more become covered by rubble, and he did not have the financial means to hire excavators to search for them. By a stroke of luck, towards the end of his time, two Glasgow engineers, Messrs Inglis and Aiton, who were working in Egypt on the construction of railways, arrived with a team of men who succeeded in uncovering all four of the great corner pieces which presumably marked the extremities of the original building. Piazzi Smyth was

A corner socket of the pyramid revealed, photograph by CPS (RSE).

delighted to be able to collaborate with them and to make use of their results. The question still remained, however, of what allowance had to be made for the marble covering layer, now missing, which clad the pyramid originally. An equally critical desideratum to establish the π theory of the pyramid was a sample of the missing casing stone which had once covered the exposed stepped structure. The angle of the pyramid's slope could be determined directly and much more accurately from an actual casing stone than by geometrical calculations based on the difficult measurement of the pyramid's height. Unfortunately most of the casing had been stripped away over the centuries and used for other purposes, but Piazzi Smyth searched and, to his great joy, found a few fragments which he brought back to Edinburgh. Their angle was later measured using special calipers constructed by Adie's to Piazzi Smyth's own design.

While tourists were a nuisance to the Smyths, especially when they were measuring the inner chambers of the pyramid, there were other more welcome visitors who came to inspect the surveying. These were engineers, of whom many were to be found in Egypt at that time, busy on a variety of construction work. Among these were the brothers John and Waynman Dixon who spent many years in Egypt engaged on the building of the bridge in Cairo and the Suez canal. After Piazzi Smyth left, they were happy to carry out further measurements on his behalf, as was also A S Grant, a Scottish doctor who was medical officer in Cairo. Some years later Waynman Dixon procured for Piazzi Smyth one of his most treasured prizes, a large intact Pyramid casing stone which was carefully measured and preserved at the Royal Observatory, and later presented to the Royal Scottish Museum. In gratitude, Piazzi Smyth lent Dixon some of his special measuring rods for the examination of other Egyptian monuments. Dixon did, in fact, make measurements on the entrance passage of the pyramid and carried out excavations in the Queen's chamber in 1872. John Dixon, an associate of Sir Benjamin Baker, was later responsible for the design of the ship which carried the obelisk Cleopatra's Needle to London in 1877.

Another engineer who was greatly impressed by Piazzi Smyth's operations at the pyramid was Max Eyth who was at the time working for Sir John Fowler, the inventor of the steam plough which had been introduced into Egypt for the cultivation of cotton. Eyth, a German, was also the author of many very popular adventure stories, of which the most famous was the *Kampf um die Cheopspyramide* (*The Battle at the Cheops Pyramid*), first published in 1902.[4] In this story there is a conflict between two brothers, one of them a worldly one, who would take away the ancient stones for modern use, the other, the romantic one with the name Joe Thinker, modelled on the character of Piazzi

Smyth, who sees in the pyramid a sacred monument enshrining the mystical π. In the notes to the book, Piazzi Smyth is mentioned by name and his mission to the pyramid explained in full. This tale became so popular that, even today, it is still in print in an abridged version and still includes the reference to Piazzi Smyth. Eyth himself was one of those who half-believed in the π theory and later in his life wrote his reflections on the subject. His delightful adventure story has, unfortunately, never been translated into English.

The Smyths' stay at the pyramid had to be limited by their finances. It is remarkable how well they achieved their target in only a few months and with a minimum of assistance. Besides their work they had many diverting experiences including a shooting in which one of their domestic helpers, a cook, was involved which resulted in his being carried off to prison in Gizeh for a time. There was also a wedding in which Jessica took a great interest, until she discovered to her amazement that the original bride was later replaced by another one without causing the romantic heartbreak which she had expected. Before they finally left their pyramid camp Piazzi arranged to photograph the families of the village, a challenging task, as the modest ladies refused to come out of their houses to be photographed

Piazzi Smyth wearing his Egyptian tarboosh (ROE).

in daylight, while indoors the babies were dandled up and down throughout the long exposures. The entire company was called to see the evening star as they had never seen it before, a dazzling elegant crescent in the telescope. 'God is great', said each one as he looked. The furniture and domestic chattels were distributed among the helpers as keepsakes, beginning, as Piazzi records, with 'magnificient globe lamps mounted in copper for the best men, descending through saucepans and frying pans, to merely miserable mousetraps for the bad subjects'. The unfortunate assistant cook who had been imprisoned requested very particularly to be given a certain tin pot 'because', he said, 'it is what they used to send me my dinner in every day when I was in prison, and I came to love it so!' For the first time the Smyths realised that unwittingly they had been feeding a prisoner!

The day came for the Smyths to take a last leave of the pyramid where they had lived and worked for four months. The helpers were paid their agreed wages, and Piazzi and Jessica set off on their camels with their paraphernalia to return to Cairo and thence home. There were tears in many eyes, and the ancient housekeeper of the tomb-home who accompanied them all the way to town declared that 'the Sheikhs and people of the Pyramid villages had not experienced anything so fully and finally satisfactory to them as the Scottish astronomer's visit'.

Chapter 7

Pyramid Facts and Fancies

Piazzi Smyth returned from his pyramid expedition with his theory not only confirmed but greatly embellished. Before going out he had had a communication from H G Haliburton, a barrister by profession who was engaged in mining ventures in Nova Scotia and was a founder member of the 'Canada First' party. He was the son of Judge Haliburton, a well-known politician and public figure who under the pseudonym of Sam Slick was the author of amusing and immensely popular tales of Nova Scotia.

H G Haliburton was also interested in anthropology and had made a study of the traditions of primitive tribes in various parts of the world. He had a theory that among many of these tribes the calendar was regulated by the Pleiades. This easily recognisable group of stars heralds the beginning of winter when it rises in the early evening and, according to Haliburton, is associated with various November festivals including Hallowe'en in Europe and festivals of the dead all over the world from India to America. Haliburton suggested that these November festivals were of great antiquity and originally denoted a year beginning when the Pleiades rose simultaneously with the Sun. On account of the precession of the equinoxes the Pleiades would indeed have coincided with the Sun in spring some thousands of years ago, when they would also rise simultaneously with the setting Sun at the autumn equinox. He was encouraged in his ideas by some scholars, including Max Müller, the German-born Oxford-based philologist. His ideas were considered by Agnes Clerke to be worth including in her discussion of the mythology of the Pleiades in the *System of the Stars*. Haliburton, who was in close touch with Piazzi Smyth all his life, though not a believer in the mystic pyramid, wrote a long essay on the subject which Piazzi Smyth reproduced as an appendix to his *Life and Work at the Great Pyramid*.

Piazzi Smyth was greatly impressed by this essay and on the voyage to Egypt he calculated the precise date when the Pleiades were in the vernal equinox and discovered that this would have occurred in 2170BC, very close to the time when according to John Herschel the star α Draconis was the pole star aligned at lower culmination with the sloping entrance passage of the pyramid. Seeing how closely the two dates matched, Piazzi Smyth added a new ingredient to his theory: that the pyramid was built when, simultaneously, the Pleiades were at the equinox and α Draconis was exactly on the opposite side of the pole. Such a configuration seemed too attractive to be ignored, and Piazzi Smyth looked upon it as part of a divine plan, which dated the building of the pyramid at exactly 2173BC. Turning his back on all archaeological evidence, which at the time tended towards a much earlier date, Piazzi Smyth became firmly convinced that the pyramid was built in that year, a date which could be fitted in with the earlier but by then mostly discredited chronology of the Bible. The Pleiades connection appeared to be clinched by an astronomical speculation which at the time aroused much interest, that our whole Milky Way system of stars was in rotation around that very same star cluster of the Pleiades.

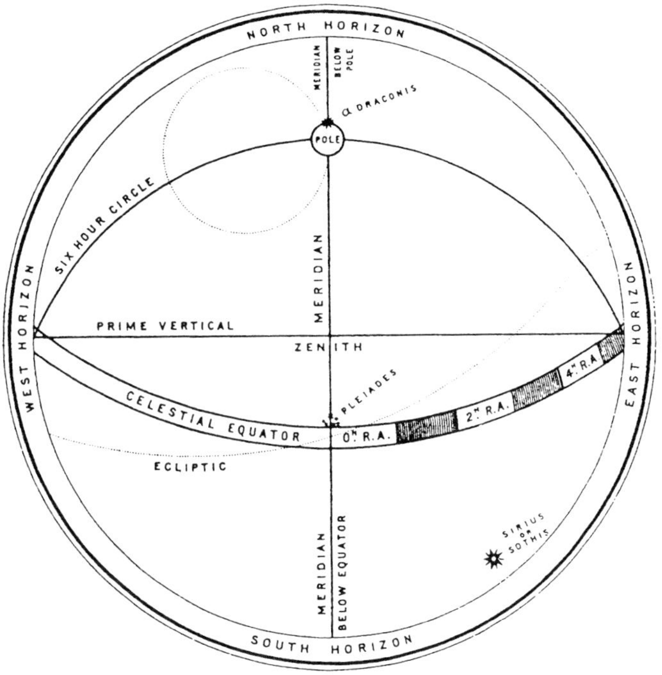

The aspect of the sky at midnight at the Spring equinox in 2173BC. Drawing by CPS (OI 3rd ed).

According to Piazzi Smyth the metrology of the pyramid had also confirmed the existence of a 'sacred cubit' of 25 'Pyramid inches', the pyramid inch being longer by one part in a thousand than the British inch, and identical with Herschel's Earth-commensurable modulus. He admitted, however, that the 'profane cubit' or the familiar cubit of Memphis could also be found in certain dimensions. In a new version of one of Taylor's 'discoveries' Piazzi Smyth also proposed that the perimeter of the pyramid contained 365.25 sacred cubits, one cubit for each day of the year, so that the pyramid proclaimed the length of the year as well as the diameter of the Earth! In an ever more astonishing argument—far removed from the scientific explanation of the lengths of the tropical and Sothic[1] or sidereal years which he had taught his Edinburgh students—he claimed that the builders of the pyramid had implanted their knowledge of the period of precession into its interior geometry. Measurements of the 'coffer' had also produced a plausible value for the density of the Earth.

The pyramid work was recorded in detail in a three-volume book *Life and Work at the Great Pyramid* which appeared two years later. The first volume was a thoroughly delightful account of Piazzi and Jessica's day-to-day experiences on their expedition. The second volume contained all the actual measurements, several thousand of them, so that if anyone so wished, he could repeat the calculations and see where Piazzi Smyth's scientific conclusions came from. This volume was an important piece of work and well worthy of

Precision clinometer made by E E Sang for Piazzi Smyth in 1869, similar to the one used at the pyramid (RMS).

publication. The same material, including exquisitely drawn diagrams, was reprinted in the *Edinburgh Astronomical Observations*. The final volume dealt with the interpretation of the results and attempted to build a self-consistent picture of the motives of the pyramid builders and the allegedly undeniable existence of astronomical knowledge enshrined in it.

The pyramid results were presented to the Royal Society of Edinburgh at the invitation of the Council in April 1866, a year after the expedition. The metrological and astronomical observations were described and interpreted in terms of Piazzi Smyth's 'sacred cubit', his 'Year of the Pleiades', his astronomical dating of the building and the commensurability of the sacred cubit with the Earth's polar axis. The 20-page paper was published as a verbatim record of Piazzi Smyth's address which, delivered in his usual style, and with the help of models and instruments, was as much a theatrical performance as a lecture. Two years later Piazzi Smyth was awarded the biennial Keith Prize and medal for this communication on his pyramid measurements. The prize was well deserved for the great care with which he had conducted the work which was accepted as an important contribution to the metrology of the pyramid until Flinders Petrie repeated the survey in 1880. The minute recommending the award, presented by the Society's President, Robert Christison, referred to 'Professor Piazzi Smyth's energy, self-sacrifice and skill', adding that 'the Council offered no opinion on the theory propounded by Professor Smyth but they were fully convinced that this was a most valuable series of measurements on which all future theories must be based'.

When Piazzi Smyth thanked the Society for the honour one of the Fellows had a few words to say 'with the utmost friendship to Professor Smyth whose measurements were most exact but whose logic was most wretched'. The Fellow was not an astronomer or a mathematician; he was a medical man, Sir James Y Simpson, Edinburgh's foremost surgeon and the discoverer of chloroform as an anaesthetic. Simpson is one of the great figures in the history of medicine, whose work is remembered in Edinburgh by the Simpson Memorial Hospital and by the contemplative statue which stands in the city's Princes Street Gardens.

This was not Simpson's first brush with Piazzi Smyth on the subject of the pyramid. Simpson was a man of many talents, a naturalist, amateur astronomer, and a serious antiquarian. He was a close friend of the Irishman Sir William Wilde, father of Oscar, himself a doctor and antiquarian and the author of a study of the megalithic burial mounds of Newgrange and Dowth entitled *Beauties of the Boyne and Blackwater*. It was Simpson's interest in this subject which brought about a debate between him and Piazzi Smyth which is recorded as one of the unforgettable episodes in the annals of the dignified Royal

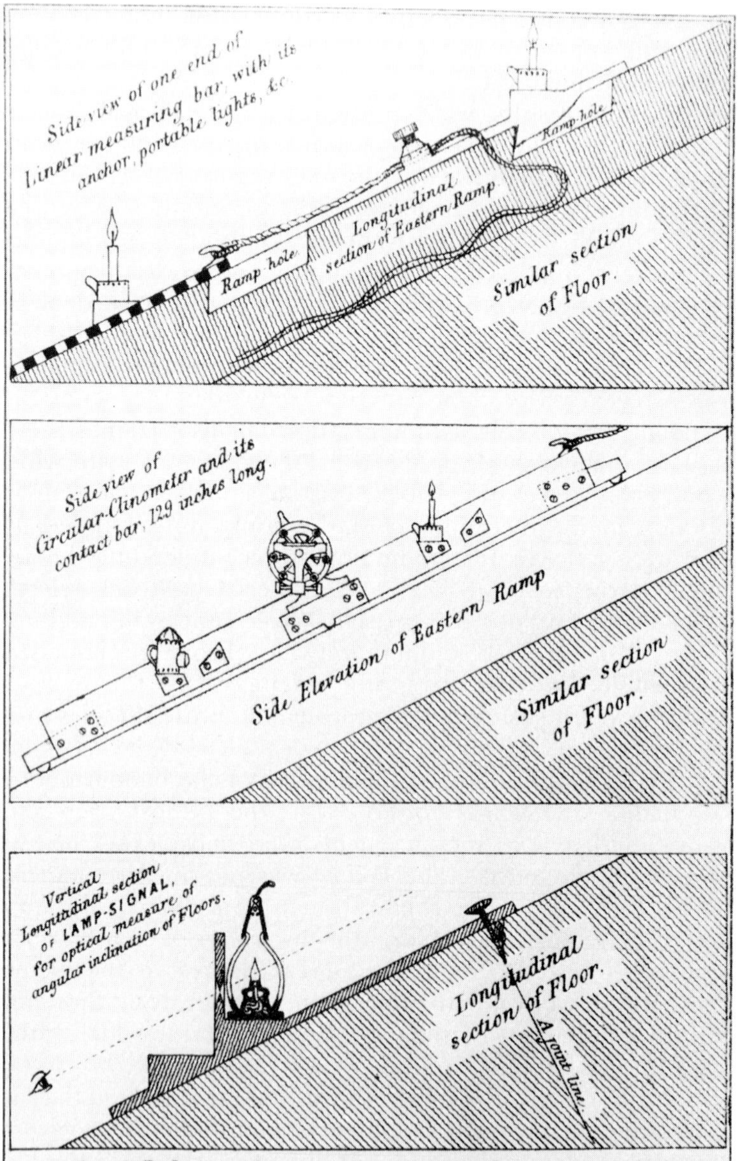

Special forms of apparatus, employed in measuring in the Grand Gallery.

Diagram showing the slopes of the inner passages of the pyramid being measured by the clinometer (EAO vol. 13).

Society of Edinburgh. Simpson, in one of his essays, had described the Irish burial mounds as 'rude but magnificent pyramids', meaning that their purpose, like that of the Egyptian monuments, was as tombs. Piazzi Smyth, outraged, refuted this remark in a paper to the Society entitled 'On the antiquity of intellectual man from a practical astronomical point of view', in which he went over the same familiar ground about the age of the pyramid and the perfection of its structure. Simpson, called upon to reply, declared that 'though Professor Piazzi Smyth's papers were extremely clever he did not believe one word of the Pyramids having been built on astronomical principles. It was a spurious archaeology and astronomy that fed such fancies and romances.' After the Keith Prize award to Piazzi Smyth, Simpson went once again into action. He took a fortnight's holiday from his medical duties in order to study the whole question, and with his amazing capacity for assimilation, he was able to go over all the historical ground, as well as over the measurements of Piazzi Smyth and of his predecessors, and to present at the very next meeting of the Royal Society of Edinburgh a paper in which he demonstrated that Piazzi Smyth's impressive conclusions were the result of selectiveness in choosing data that suited him and in taking averages of divergent numbers ending up with the supposed exact result. The delivery of this lecture was accompanied by an amusing demonstration which was reported in the newspapers: 'The lecturer stated inter alia that if Sir Isaac Newton's standard of the sacred cubit was true, and if the polar axis of the Earth be, as fondly imagined by Professor Smyth, 500 million pyramid inches, then the brim of the lecturer's hat being exactly one half of Sir Isaac Newton's sacred cubit, that brim was in measurement one 20 millionth of the Earth's polar axis.' The hat was produced and the measurement performed by an assistant with—as Piazzi Smyth afterwards described it—'so much unction of manner and look as to be received with cheers by the large and learned audience in the hall of the RSE'. What was ironical about the particular criticism made by Simpson was that it attacked the aspect of his theory where Piazzi Smyth felt he was on the strongest ground, since Sir John Herschel himself had been the one who highlighted the commensurability of the cubit and the length of the Earth's polar axis. There is no doubt that Simpson was right and that Herschel had been unwise in claiming significance for this coincidence.

Piazzi Smyth's reaction to Simpson's discourse was another lecture, which in its printed form occupied 60 pages in the *Proceedings of the Royal Society of Edinburgh*, in which every accusation was rebutted point by point, to his own satisfaction, in lofty tones, accompanied by impressive references to ancient history and biblical authorities. In this paper the revealed religious aspect of the pyramid theory was unequivocally stated. He ended his discourse by comparing himself with Kepler:

> The existing state of our Great Pyramid knowledge is very similar to what prevailed with regard to astronomy in the time of that acutest genius of mediaeval Germany; for observations were then rough, and rude hypotheses succeeded each other as found from time to time best adapted to explain all the known facts, and were not worked up into a complete theory. So many failures had at that time already taken place that when success did come at last, hardly any one would believe Kepler's brilliant discovery of the elliptical motions of the planets. His friends would hardly allow that he was sound in his mind; yet his discovery proved the morning star of the epoch of the Law of Gravitation. They ridiculed and despised: but thus he comforted himself in a knowledge of a reality, and a presentiment of the importance of what he had learned; and with an intellectual enthusiasm not often seen in the present day of a more delicate education, he thus breathes forth his noble spirit: 'Nothing holds me. I will indulge my sacred fury! If you forgive me, I rejoice; if you are angry, I can bear it. The die is cast. The book is written to be read either now or by posterity, I care not which. It may well wait a century for a reader, since God has waited 6000 years for an observer!'

This paragraph sums up Piazzi Smyth's strange delusion of the pyramid as a repository of scientific knowledge, purposely hidden until man was sufficiently advanced in science to appreciate its truth. In a contorted way he had the unshakeable belief in progress which was characteristic of his times, and the conviction that there was no limit to the power of science.

The Simpson controversy continued, with further insults exchanged between the protagonists. It was strange that the Royal Society of Edinburgh allowed Piazzi Smyth a platform for his plainly non-scientific ideas, and permitted the unseemly quarrel to drag on. Piazzi Smyth protested to the Council when Simpson's scathing comments were actually printed in the *Proceedings* of the Society. The Council replied with a conciliatory letter but Piazzi Smyth, unappeased, declared his intention to return the Keith Prize and to 'cease to subscribe himself FRSE'.

The Council of the Royal Society of Edinburgh reacted to this outburst with great tact. The President went in person to Piazzi Smyth's home and persuaded him to withdraw his threats. This Piazzi Smyth gladly did, in a letter in which he declared that he would 'show in future as far as my poor exertion can, that I regard the health, strength and independence of the RSE as of the highest importance to the progress of science not only in Scotland but in Great Britain and the world'. Following a further complaint the

Council responded in the same gentle manner, saying that they could not act as '*Censor mororum or Censor litterarum*' but that 'anyone acquainted with Professor Piazzi Smyth as they were themselves could not fail, whether they agreed with him or not, to treat him with respect and do justice to his conscientious diligence, his perfect integrity as a scientific enquirer and his desire to conduct controversy with the courtesy of a gentleman'.

Around this time Piazzi Smyth faced another attack on his π theory of the pyramid. It came from his friend Sir Henry James, Director of the Ordnance Survey, whose engineers had recently carried out a survey of Egypt. James had looked forward with great interest to Piazzi Smyth's metrological results at the pyramid and when he received Piazzi's book he was prompted to examine for himself the measurements he found there in the light of his own survey. He discovered that the corners of the pyramid rise towards its apex in the proportion of nine parts in ten, which meant that 'for the construction of the Pyramid the architect therefore had to set up profiles of the four angles which had a vertical height of 9 feet or 9 cubits to a horizontal length of 10 feet or 10 cubits, for such profiles would be sufficient to guide the masons at every stage of their work from the base to the apex'. This discovery was a terrible blow to Piazzi Smyth's ideas, because such a model produces a pyramid with an angle of slope differing by only half a minute of arc from that of the perfect π model, and linear measures differing by only one part in 3000, differences far too small to be detected by his measurements which he had earlier claimed could give the slope angle correct to one minute of arc and the linear dimensions to 1 part in a 1000. (In fact his measurements were much less accurate than this; his published results show that the uncertainty was several times greater than he had first estimated.)[2]

Realising that James' simple explanation could not be refuted from the observations, Piazzi Smyth was driven to fall back on the motives of the builders, thus taking the whole argument out of the realm of such science as was there. An old friend and loyal supporter came quickly to his aid. This was William Petrie, his companion from South African days, now an engineer practising in London who kept up an interest in all Piazzi Smyth's work. Petrie, and more especially his wife, were devout Christians who accepted the divine origin theory of the pyramid. In letters to the *Athenaeum* Petrie tried to reconcile the theories of 'the Astronomer and savant with the Scriptures reverently in hand; and the gallant Director of Military engineers surveying en masse the whole of the country' by saying that, though the 9:10 ratio might well have been given to the workmen as a guide, the intention had been to construct a π-pyramid. Furthermore—and here was a new and marvellous idea—the numbers 9 and 10 revealed something more, namely the distance from the Earth to the Sun which is 10^9 times the height of the pyramid.

Petrie expounded this view in a paper submitted through Piazzi Smyth to the Royal Society of Edinburgh. He quoted the best astronomically determined values of the Earth–Sun distance to demonstrate how beautifully it fitted into the dimensions of the pyramid. The referee, Kelland, reported that 'the object of the paper obviously is to prove the existence of supernatural revelations which lie entirely outside the domain of science'. The rejected paper appeared later as one of the innumerable pamphlets which Piazzi Smyth distributed among his faithful followers.

This and other wondrous revelations were included in Piazzi Smyth's second popular book, *The Antiquity of Intellectual Man*, written in 1868 as a sequel to *Our Inheritance in the Great Pyramid*. The title suggests that the book was intended as a challenge to Sir Charles Lyell's famous *The Antiquity of Man*, published in 1863, which described the geological evidence for the theory of evolution, a question which was at the time widely debated and was a stumbling block to many devout people. The book displays much erudition in the form of references to geological and historical sources, but, as always, these are not discussed except to disagree if they do not fit in with Piazzi's own views. 'Though whole shiploads of chipped flints may be dug out of any number of gravel beds and may be claimed by anthropologists as contemporary remains of those who shaped them ages ago, they testify nothing to intellectual man'; it was the old argument that the pyramid could not have been designed by contemporary people, and that was all there was to be said on the matter. One wonders what his wife Jessica felt, who in her girlhood had filled her geology notebook with accounts of the latest discoveries in evolution; and at the change in the attitude of Piazzi, who had recorded his amusement at the simple-minded anti-evolutionist whom he met during his visit to Russia not many years before.

Piazzi Smyth's father who had been in poor health for a long time, died in September 1865, a few months after Piazzi and Jessica's return from Egypt, and long before the publication of the results and the ever more fantastic interpretations of them. The wise old Admiral, had he lived, might well have restrained his son's post-expedition excesses. Perhaps, too, if Herschel had spoken, Piazzi Smyth might have desisted. Piazzi was under the delusion that, if his metrology could establish the π formula in the pyramid design, Herschel would be obliged to support his theory with all its implications. He was never to have that satisfaction. Herschel's name was brought into every one of Piazzi Smyth's pyramid writings; but Herschel, a man who avoided all scientific feuds, kept his counsel.

Life and Work at the Great Pyramid was favourably received by some of the press, including such highly respected newspapers and journals as the *Illustrated London News*, the *Standard*, the *Scotsman* and the *North British Review*. 'Professor Piazzi Smyth deserves no common credit for his remarkable investigations; and we think he has made out a case for further enquiry. This, however, ought not to be left to private means but developed by the resources of a powerful State.' was one comment. 'The Egyptians upwards of 4000 or 5000 years ago must have possessed a complete metrological system, more perfect than the metrical system of which we, the French, are so proud, and they have engraved the indelible remembrance of it on this mountain of stone', said the French *Revue des Deux Mondes*. The *Daily Review* best summed up the spirit and attraction of Piazzi Smyth's hard work and enthusiasm which shine through the book: 'As a record of pious devotion to science, and of labours undertaken from the loftiest motive by which man can be inspired—a love of truth for the sake of it and a determination to get at it—these magnificent volumes will ever have a moral charm. His conclusions it is difficult to accept. It is difficult also to reject them.'

Difficult to accept; difficult to reject. As in some modern 'findings' in astrology not everyone could spot the flaws. Viewed in the context of the ideas prevailing in mid-nineteenth-century Protestant Britain where a high proportion of the population was Bible-centred in religious matters, they were not so utterly incredible. The original theory made just two claims: that the pyramid was built on the π pattern, and that it was laid out in units of the sacred cubit, a measure of biblical antiquity which had been sanctioned by Isaac Newton himself in his 'Dissertation on Cubits' and which now somehow also had the authority of the great Sir John Herschel. These two facts, scientifically proved according to Piazzi Smyth, led to the idea of divine inspiration, and among many of the general public, this was the really attractive part of the theory. Here, it seemed, was an actually existing monument which could be a direct link with the sacred edifices of the Old Testament. Popular books on all aspects of nature—astronomy, geology, botany—which were published under the aegis of religious organisations like the Society for Promoting Christian Knowledge usually invoked the Creator and looked upon the discoveries of science as revealing God's handiwork. Piazzi Smyth's father's immensely popular *Cycle of Celestial Objects* was itself a good example of this attitude: 'It is in the heavens that the sovereign wisdom manifests order and harmony, regularity and proportion, greatness and majesty, arrangement and destination, in astounding immensity.' In America, O M Mitchell's book, *Orbs of Heaven*,[3] based on lectures in Cincinnati in 1843 which created such an impression that the audience was stirred into raising the funds for the country's first substantial telescope, actually began with Adam and Eve's reactions to the first day of creation.

Piazzi Smyth had been brought up in this climate, and his religious convictions would never have caused him to be condemned, even in educated circles. It is therefore not surprising that he was not without supporters among the older generation of scientists who accepted his theories in varying degrees. His friend William Petrie was the first of these. Another was Hamilton L Smith of Hobart College, Geneva, New York State, who had set up a small Department of Astronomy in what was then a private college. He was about the same age as Piazzi Smyth and they came to correspond through their common interest in photography. After the Egyptian expedition Piazzi Smyth sent him a copy of his paper on photography at the pyramid, which led to Smith taking a keen interest in looking for apparently significant relationships between the various dimensions of the pyramid, which he communicated to Piazzi Smyth in letters extending over many years. His investigations, described in a long paper published in 1873, supported the view that the pyramid builders possessed a high—though not superhuman—degree of astronomical and geometrical knowledge. Joseph Baxendall, a well regarded amateur astronomer who was for a time Astronomer to the Corporation of Manchester, was similarily impressed by the coincidences of the pyramid's geometry.

Henry Mitchell of the US Coast Survey who belonged to a surveying team in Egypt in 1868 was also a sympathiser. He had noticed—as Napoleon's scientists had done before him—that the pyramids of Gizeh occupied a central position with respect to the circular outline of Egypt's Mediterranean coast. He supplied Piazzi Smyth with material which, according to Piazzi Smyth, clearly showed that the pyramid was in fact in a unique position in the world, in the very centre of the globe's landmass. Mitchell and his wife appear to have visited the Smyths in 1868 when they no doubt recalled the visit which Henry Mitchell's sister Maria had paid to Edinburgh shortly after the Tenerife expedition when she had met Robert Stephenson, the engineer, and Robert Chambers, author of *Vestiges of Creation*, a forerunner of later theories of geological evolution.[4] Maria Mitchell, Professor of Astronomy in Vassar College, was the world's first professional woman astronomer.

Another strong supporter of Piazzi Smyth's theories was the French astronomer and mathematician, Abbé F Moigno, who was the founder and editor of a respected journal called *Cosmos*, later renamed *Les Mondes*, which published papers and carried reports of scientific meetings. Moigno was a Catholic priest who fell out of favour with his immediate superiors and was for a time banished to teach in a remote school before he returned to Paris where he founded his journal and served as a curate in a village outside the city. Piazzi Smyth first met him at the Paris Exposition of 1855; they met again in 1871 at the Edinburgh meeting of the British Association for the Advancement of Science, which is probably the time when Piazzi Smyth converted him to his ideas on the pyramid. Piazzi Smyth was an evangelical with an antipathy to Catholics and therefore theirs was an extraordinary friendship; and an unexpected one on other grounds, too, because Moigno was an early advocate of the metric system and a member of the international committee for the universal adoption of the metre. What may have contributed to Moigno's attraction to Piazzi Smyth's ideas was the fact that Piazzi Smyth was so devout a Christian in an increasingly agnostic age. Moigno promoted the pyramid theory in his journal and translated an abridged version of *Our Inheritance in the Great Pyramid* into French which was published in 1875. Moigno's English was poor—he used to write to the Smyths in French—but the translation was a labour of love for Piazzi and Jessica who visited him whenever they came to Paris. The publicity given to the pyramid in *Les Mondes* was responsible for spreading the cult to France and Belgium, though Piazzi Smyth does not appear to have entered into direct correspondence with any of his continental followers.

Piazzi Smyth had another scientific outlet in the *Quarterly Journal of Science* where he published in 1871 a long article containing all his scientific and historical work on the pyramid. This journal was founded and edited by William (later Sir William) Crookes, a scientist of wide-ranging talents, chemist, physicist, spectroscopist, but most celebrated for his discoveries in radioactivity. He was a man of independent mind who considered all phenomena, however odd, to be worth investigating and whose own experiments in psychical research did not meet with general approval. He was happy to publish Piazzi Smyth's unconventional researches and to allow him considerable space to review a book on *The Great Pyramid of Gizeh* by A F D Wackerbarth, a mathematician who had attacked Piazzi Smyth's metrology.

Much interest was shown on strictly scientific grounds in Piazzi Smyth's observations by the geologist Robert Mallet who hoped that they might provide him with useful data in his researches into the Earth's crust and the transmission of earth tremors by his postulated 'waves of compression'. He suggested that the slight discrepancy between the measured azimuth of the pyramid's face and the true east–west line might have arisen through such waves. Piazzi Smyth favoured this explanation in *Life and Work*. Mallet also considered that any evidence that the pyramid had sunk or become distorted since it was erected would provide quantitative information regarding the compressibility of the Earth's surface material.[5] Another geophysicist, Antoine d'Abbadie, suggested that an observation of the deflection of the plumbline by the pyramid on the principle of Maskelyne's experiment at Schiehallion could be used to find the mass of the Earth in terms of the mass of the pyramid which could be accurately calculated from Piazzi Smyth's data.

On the other hand there were scientific friends who were uneasy. Amongst them was E B Denison (Lord Grimthorpe) who had a common professional interest in clocks. He expressed disappointment at Piazzi's advocacy of the 25-inch cubit and offered a very obvious solution to the metrology, which 'stared one in the face', namely, that, if instead of using the 'sacred' 25-inch cubit Piazzi Smyth had tried the normal Egyptian cubit of 20 inches, he would have realised that 440 of these made up the side and 280 of them the height of the pyramid, thus giving 22/7 for the relevant ratio. It is interesting that Piazzi Smyth always refused to consider this close approximation to the number π, which would have put a simple pair of numbers in place of a mystic figure. In a popular book, *Astronomy without Mathematics*,[6] Denison disagreed with Piazzi Smyth's theory and declared that the pyramid was no doubt a perfectly straightforward case where builders had used rods to make their measurements.

Fragments of casing stones brought back by Piazzi Smyth from the pyramid and mounted by him in a model of a perfect π-pyramid (RMS).

Fifteen years later the very same uncomplicated explanation for the dimensions of the pyramid was to strike Flinders Petrie.

When *Life and Work at the Great Pyramid* came out, Jessica began to keep a scrap-book into which she pasted newspaper comments, reviews and letters, which show that many reputable scientists were attracted to one or other aspect of the pyramid results. The suggestive astronomical elements were exciting; R S Proctor, the highly respected writer and populariser of astronomy who was later to disagree with Piazzi Smyth on the subject wrote: 'Your labours have established results which seem to me to be as interesting and valuable in their own way as those established by your experiments at Tenerife', a remark that pleased Piazzi Smyth so much that he included a flattering reference to Proctor in the next edition of his *Our Inheritance*. Balfour Stewart's comment was 'I think there can be little doubt that the Great Pyramid is a standing memorial to the astronomical knowledge of the ancients.' Piazzi Smyth's friend Joseph Sidebotham, besides arranging to have the Egyptian photographs copied and distributed, contributed a paper to the Manchester Literary and Historical Society in which he described Piazzi Smyth's flash photography in the interior of the pyramid and drew attention to his method of including scaled rulers in the photographs as a means of demonstrating the dimensions of the structure. Sidebotham became a devoted pyramidologist. 'I like the tone of your book', he said, 'your recognition of the Creator as Governor of the world.' James Nasmyth, that remarkable engineer–astronomer–photographer was also most impressed: 'The subject is in every respect one of the deepest interest and has been, by your admirable researches, invested with a character which places it beyond all merely antiquarian objects.' Nasmyth, however, was on more familiar ground when he discussed Piazzi Smyth's photographic achievements at the pyramid. The Rev. Alexander Mackay, amateur geologist and author of school textbooks on geography, a contact from Jessica's youth, incorporated a special chapter on the pyramid in an edition of his textbook and gave lectures on the subject in and around Edinburgh. Another enthusiastic Edinburgh clergyman was the Rev. Joseph Goodsir, a member of a fraternity called the 'Brotherhood of Truth'[7] which was founded in the 1830s by a group of students in the University of Edinburgh, many of whom became leading academics in later life. It is significant, however, that no account of the pyramid appeared in *Good Words*, the widely read liberal evangelical review edited by the Rev. Norman Macleod, Moderator of the Church of Scotland in 1869–70, to which such scientists as Brewster, Herschel, Forbes, Thomson and Tait contributed popular scientific articles. Piazzi Smyth had contributed four long illustrated articles to *Good Words* in the two years before his

Egyptian expedition but none afterwards; it is not unlikely that Norman Macleod excluded Piazzi Smyth's pyramidology from the ranks of genuine science.

Quite a different aspect of the pyramid theory which aroused interest in academic quarters was the chronological one. The rapid progress in Egyptian archaeology which was flourishing throughout the nineteenth century promised the possibility that archaeological remains which could be associated with the Israelite sojourn in Egypt might be discovered and dated. Henry Smith Williams, author of the article on chronology in the *Encyclopaedia Britannica* of 1911 says:

> Considering the important part played by the Egyptian sojourn of the Hebrews as narrated in the Scriptures, it was certainly not an over-enthusiastic prediction that the Egyptian monuments when fully investigated would divulge important references to Joseph, to Moses, and to the all-important incidents of the Exodus; but half a century of expectant attention in this direction has led only to disappointment.

It was the claim made by Piazzi Smyth that the pyramid was associated with the Hebrews while in Egypt which aroused the interest of some scholars such as the members of the Society of Biblical Archaeology, founded in 1870, which Piazzi Smyth was invited to join by Samuel Birch of the British Museum, an old friend of his father and of Lee. He was also asked to take part in an Oriental Congress in 1874 but he does not appear to have accepted. Birch did not personally believe 'that the Pyramid in its original destination was otherwise than a monarch's sepulchre and a nation's pride', but the fact that he invited Piazzi Smyth to join the Society indicates that the latter's work was by no means considered unacceptable. Piazzi Smyth preferred, unfortunately, to associate with less critical admirers who regarded his doctrines as proven scientific facts.

Piazzi Smyth distributed his publications widely. Copies of his scientific papers, which included those on pyramid metrology, reached all scientists on the Royal Observatory's mailing list and were also sent to anyone who showed a glimmer of interest. Piazzi gave away his own pyramid books at considerable expense to himself and distributed simpler pamphlets or reprints from various magazines to general enquirers. He never appeared personally at religious meetings but rather supplied material to would-be authors which would later appear as articles by a number of independent pamphleteers, when in fact much of the activity was masterminded from Edinburgh. On looking through the surviving pyramid correspondence one must marvel at the energy and efficiency with which the operations were carried out.

An important aid to the spread of the pyramid doctrine were Piazzi

Smyth's splendid Egyptian photographs which were available in the form of lantern slides. These were produced by the commercial photographer J S Pollitt in Manchester, whom Piazzi Smyth met through his friend Sidebotham. Pollitt's advertisement in Piazzi Smyth's books read:

> J. S. Pollitt has pleasure in announcing that by the kind permission of Professor Piazzi Smyth he is now publishing this fine series of pictures printed on glass and suitable for use in the magic lantern. At the urgent request of Pyramid students enlarged positives are also being printed on paper. The rare excellence of these pictures and their suitability for Illustrating Lectures on the Pyramids and more especially on the Scientific Construction of the Great Pyramid as an august primeval monument of number, weight and measure, have long been known and appreciated by those who have seen them.

'Pyramid studies' had become a new branch of learning, or at least a hobby which gave pleasure to many people. One of the earliest converts, a clergyman, kept up his lecturing for the remainder of his life, and in 1883 at the age of 86 could write rather charmingly and naively to his mentors:

> I often think it was a most auspicious moment when I casually stumbled on your first edition of 'Our Inheritance'. From that moment I have felt the deepest interest in the subject, as deep as a simple ignoramus could take, being utterly ignorant of the first principles of mathematics and not even able to spell the word logarithm, and as for the value of π, it far exceeds my comprehension. To you yourself and the equally kind-hearted Mrs P. I have been most indebted for the great pleasure of having a thorough conviction stamped on my mind of the truth of your noble and solemn theory of the glorious old Pyramid.

Unfortunately, by the time the pyramid doctrine had spread to this extent, after several years of input by various busy disciples, it had moved a long way from the first simple ideas of metrology and geometry. The exercises in number-juggling by which Piazzi Smyth managed to extract the period of precession and the Earth–Sun distance from the pyramid measurements gave the signal to his more numerate disciples to play games with all the numbers associated with the pyramid—dimensions, volumes, not to mention the magic number π itself—in a search for clues to new scientific information. Extraordinary coincidences between these kept turning up, many of which impressed Piazzi Smyth himself, though they were no more than circular calculations or the result of the chance circumstance that π squared is close to 10 which makes the inverse of π numerically similar to π itself. If so much had already been discovered, so the devout disciples thought, surely there was more to impart; and, if the pyramid contained all scientific knowledge up to the present time, it could be but a short step to demand, as Piazzi Smyth's more enthusiastic followers did, that it should also disclose new, as yet unrevealed, facts about the world. It was at this stage that Piazzi Smyth's pyramid movement began to go badly wrong, with fanatical religious followers, who unfortunately had an influence on their master, looking for fulfilments of obscure biblical prophesies. The matter of chronology also took on a serious aspect, with attempts to reconcile the dating of the pyramid with Noah's flood and with the biblical chronology of Archbishop Ussher who had placed the creation of the world at 4004BC. Piazzi Smyth's fundamentalist stand on this matter contrasts unhappily with the wholly scientific views he had held in Russia, when he joked with Otto Struve about the notion that Noah's Ark was lying on the summit of Mount Ararat. Not a few of Piazzi Smyth's well-wishing friends, among them the clergyman and astronomer Henry Cooper Key, warned him of the danger of listening to what one of them called the 'claptrap' of bogus scholars and 'charlatans who pepper their nonsense with Scripture quotations'; but the pious Smyths continued to be drawn into these unfamiliar waters.

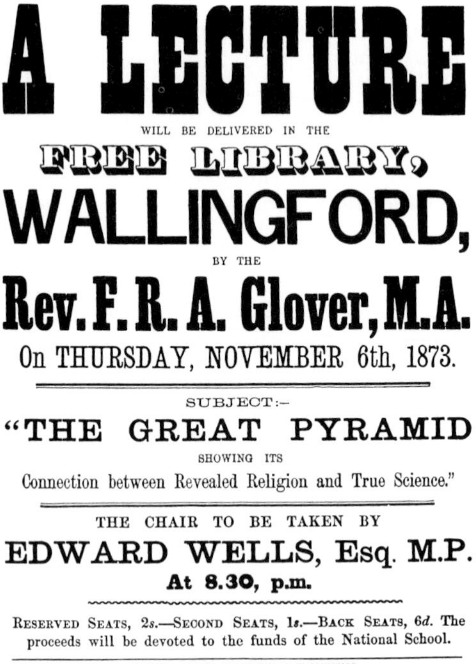
A LECTURE
WILL BE DELIVERED IN THE
FREE LIBRARY,
WALLINGFORD,
BY THE
Rev. F. R. A. Glover, M.A.
On THURSDAY, NOVEMBER 6th, 1873.

SUBJECT:—
"THE GREAT PYRAMID
SHOWING ITS
Connection between Revealed Religion and True Science."

THE CHAIR TO BE TAKEN BY
EDWARD WELLS, Esq. M.P.
At 8.30, p.m.

RESERVED SEATS, 2s.—SECOND SEATS, 1s.—BACK SEATS, 6d. The proceeds will be devoted to the funds of the National School.

S. BRADFORD, PRINTER AND STATIONER, WALLINGFORD.

A poster advertising a lecture on pyramidology (ROE).

A large following of another kind was attracted to the pyramid as the result of Piazzi Smyth's ideas which connected the lengths of the British inch and the sacred cubit. These were the adherents of the British-Israelite movement who believed that the British nation, or rather the Anglo-Saxon races, were in fact descended from one of the lost tribes of Israel. Since Piazzi Smyth's championing of the inch against the French metre was at the root of his pyramid theories, he could hardly turn away from this new source of support. In logic, if the inch was of Hebrew origin, so too must be the people who preserved it. Piazzi Smyth allowed himself to be drawn into discussion with these zealots to the extent that in the third volume of *Life and Work at the Great Pyramid*, published in 1868, he devoted several pages to the writings of this sect and extolled the superior degree of civilisation of the Anglo-Saxon races in a manner quite alien to his true personality. A wise friend, Henry Yule, the geographer and oriental scholar, chided him in strong terms; such direct references to the superiority of the Anglo-Saxon race did not appear again in any of Piazzi Smyth's writings, though he continued to be associated with the British-Israelite organisation, writing periodically in their magazines and ensuring in this way further circulation for his own ideas. In 1872 Piazzi Smyth was invited to take part in a conference organised by this movement in London, and was actually put on the Council of the Conference for that year, in the company of a list of impressive-sounding names which included a bishop, four clergymen and three medical doctors. His contribution, which was read on his behalf, was 'Shall we have false, or true Earth-commensurable Measures?'; it was reprinted in pamphlet form and widely distributed. It should be mentioned that William Petrie, in all other aspects a staunch supporter, did not share Piazzi Smyth's sympathy with the views of this movement.

The later editions of *Our Inheritance in the Great Pyramid* show how deeply Piazzi Smyth had become trapped in strange fields which overlaid and virtually replaced his original simple 'astronomical' theory. References were no longer to scholarly works of history or archaeology but to pamphlets emanating from the brotherhood of zealots. Five years of the life of an able and imaginative experimental scientist were lost on this fruitless cause, and even afterwards a large fraction of his time for the next 20 years was wasted. Only someone with his amazing energy could have carried out his serious and impressive scientific work under the burden of silly correspondence with utter cranks. Piazzi Smyth's influence in the cause of the mystic pyramid was enormous. In his lifetime the cult had spread in Britain, America and the distant parts of the British Empire where it was carried by zealous clergymen. To a lesser extent it had some following on the continent of Europe through the influence of Moigno. A century later the reverberations could be still felt. A history of pyramidology by the nephew of one of Piazzi Smyth's disciples, written as recently as 1971, includes a dissertation attributing mathematical properties to the pyramid far more elaborate than anything dreamt of by Piazzi Smyth, and a bibliography containing dozens of titles of writings on the mystical and prophetic aspect of the subject right up to the date of publication.[8]

Augustus de Morgan, the urbane mathematician and old friend of Piazzi's father, published in 1872 a wholly delightful compilation of unusual ideas and fallacies under the title *A Budget of Paradoxes*,[9] in which he gave some prominence to Piazzi Smyth's fantastic theory of the pyramid—a paradox indeed if ever there was one.

Chapter 8

Munich Opticians and a Bavarian Astronomer Royal

Piazzi Smyth's preoccupation with the pyramid lasted almost five years. In that time the observatory Board of Visitors met only once, in January 1867, when Piazzi Smyth called a special meeting to discuss the problem of accommodation and to urge that one of the large houses which were then being built in Edinburgh's Royal Terrace on the north slope of Calton Hill be bought as a combined residence for himself and offices for the observatory. The suggestion was agreed to and a request for the necessary money was sent to Parliament where all matters Scottish had a vigorous champion in Duncan Maclaren, by then a Member of Parliament. Thanks to Maclaren, a sum of £4000 was voted for the purchase of the house and R Matheson, the architect of the Office of Works, was instructed to supervise the necessary alterations to make the house suitable for the observatory's purposes.

At their next meeting in April 1869 the members of the Board reiterated their warm support for Piazzi Smyth's long-standing request for a good equatorial telescope. He almost spoiled his case, however, when he asked at the same meeting for permission to return to Egypt 'with the purpose of obtaining at the expense of the Nation more trustworthy measurements of the Great Pyramid'. The Treasury was, understandably, not prepared to accede to this latter request but was willing, at last, to consider the purchase of the telescope. The Treasury put the telescope into their estimates for 1871, but Piazzi Smyth first needed to obtain as much information as possible about telescopes available at the time. Being like other astronomers as yet undecided on the relative merits of lens and mirror telescopes, refractors and reflectors, he reached the conclusion that he should first visit the leading optical establishments on the continent of Europe and particularly in Munich which had produced first-class refractors such as the 15-inch for Pulkova and were constructing a 13-inch telescope for Greenwich.

In May 1869 Piazzi Smyth and his wife set out for Munich, calling in London for a few days to consult instrument makers such as John Browning, the optician, on whose work Piazzi Smyth was later to depend considerably, and Louis Casella who had just made for him some finely graduated 'Standard Scales of British Inches'. There were friends to be looked up, among them Joseph Bonomi, Curator of the Sir John Soane Museum and an old friend of Piazzi's father and of Lee of Hartwell. Bonomi's name was closely associated with Egyptian antiquities; he had spent eight years in Egypt, including some time with the great Lepsius, and was the illustrator of many works on Egyptology. At the Soane Museum the Smyths also met William Simpson, the artist, who had just returned from the pyramid and could discourse with Piazzi on his favourite topic. It gave Piazzi particular pleasure to have time to spend with his old friend William Petrie, who had never ceased to believe in the secrets of the pyramid, and his wife and 16-year-old son Flinders whose mastery of Egyptology was already evident when he conducted the Smyths around the British Museum. There Piazzi Smyth and Flinders met Samuel Birch, Keeper of Oriental Antiquities, and were permitted to make measurements of the three Great Pyramid casing-stones which had been donated to the Museum by Howard Vyse. One evening the Petries and the Smyths attended together a meeting of the Plymouth Brethren in the suburb of Lewisham where Petrie was the main speaker and where there was 'much Pyramid talk, prayer and Scripture expounding'. The Smyths stayed with Piazzi's mother, who was now settled in London, and on the last evening the family—Piazzi's brother Warington and his wife, his sisters Ellen and Rosetta and their husbands—gathered at her home in Inverness Road. The family were careful not to raise the topic of the pyramid. Piazzi was annoyed at this. 'No satisfactory talk about either family affairs or Pyramid; not one word about the latter the whole time'. He was in a disgruntled mood when he and Jessica left London docks next morning for Antwerp and their journey into Germany.

Their first stop was Bonn where Piazzi Smyth was eager to meet F W Argelander who, between 1852 and 1859, had been engaged on his marathon *Bonner Durchmusterung*, a catalogue giving the positions and magnitudes of some 320 000 stars down to ninth magnitude. This 'B.D.' had been produced by Argelander and two collaborators, E Schönfeld and A Krüger, using a small visual telescope of only three inches aperture. The Smyths found Argelander at the observatory, a striking building in the tree-lined Popplesdorfer Allee which had been completed in 1845 under the royal patronage of King Friedrich Wilhelm IV with its one great dome and four smaller ones, standing in a delightful garden. Piazzi Smyth was much impressed when he saw the tiny telescope and thought of the massive catalogue of stars which had been obtained with it. He inspected other instruments but noticed their imperfect state, due to the observatory's limited resources. On the subject of telescopes and telescope-makers Argelander favoured the Munich opticians Steinheil as producers of excellent long-focus object glasses, superior, he said, to those of Merz. Argelander, now 70 years old, and taken up entirely with his catalogue was, however, 'not likely to go out of his way for anyone'.

If the meeting with Argelander had been somewhat unenthusiastic, the warm reception which the Smyths received from another astronomer, J H Mädler, was all the more appreciated. Mädler, then 85 years of age, had retired to Bonn from a long term in Dorpat in Estonia where he had succeeded the elder Struve as Director in 1840. Before this he had worked at the private observatory in Berlin of W Beer, a wealthy banker, where in 1837, Beer and Mädler had published their compendium on the Moon, the *Mappa Selenographica*, which made a major contribution to selenography. Piazzi Smyth was disconcerted to find that Mädler had somehow fallen out with the Struve family in the meantime and was particularly critical of Otto Struve who had been the Smyths' generous host in Pulkova ten years earlier. However, Piazzi Smyth and Mädler had one important interest in common—the exploration of the motions of the stars in space. In 1846 Mädler had published a paper in which he claimed that the whole universe of stars revolves around a centre which he placed in the region of Alcyone, the brightest star in the cluster of the Pleiades. This was a suggestion which Sir John Herschel in his *Outlines of Astronomy* had considered 'utterly improbable' because the Pleiades lie so far off the circle of the Milky Way. Though by the 1860s most astronomers had come to the same view, Piazzi Smyth was not prepared to reject Mädler's idea out of hand. The central Pleiades seemed to fit markedly well into his ideas about the history of the Pyramid which, according to him, had been built when the Pleiades were in the vernal equinox. Mädler and Piazzi found, therefore, that they had much to talk about. The only mishap was that during dinner at the local beauty spot of Godesberg on the Rhine the old astronomer fainted; as Frau Mädler explained, he was 'over excited thinking of the Pyramid stars'.

After Bonn the Smyths continued their journey to Munich travelling first by steamer up the Rhine to Bingen and from there by rail to the capital of Bavaria. In spite of the scenic beauties and historical associations of the famous route Piazzi Smyth's German notebook displays little of the gusto which marked his accounts of his tours in Russia and Egypt. Of the unequalled scenery of the Rhine and the Bavarian mountains which Piazzi in his normal mood would have found irresistible there is not a single sketch or even a verbal description. At this stage in his life the influence of his pyramidologist acquaintances seems to have coloured his reactions to places as well as to people. Travelling through predominantly Catholic surroundings Piazzi filled page after page of his diary with religious comments of a disturbing and quite uncharacteristic narrow-mindedness.

However, once the travellers arrived in Munich—where they stayed in the Hotel Vier Jahreszeiten (Four Seasons), 'a splendid palace of a building, abundance of waiting, well managed and looked after'—they found it impossible to resist the attractions of the beautiful capital. Piazzi enthused over the wonders which had been worked by the late King Ludwig I and philosophised on the genius of the Bavarian mountain people. He admired the shops, where even the grocers and haberdashers had their windows artistically decorated and where the book, print and other shops 'for the mind' were remarkable for their numbers. The Smyths toured the museums and art galleries, explored the Hall of Fame and went out to Nymphenburg, Munich's Versailles with its palace, park, fountains and statuary. They visited artists' studios and purchased paintings. Though they deplored Catholic services they visited the twin-domed cathedral, the Frauenkirche, and the famous churches of St Ludwig and St Michael in the Ludwigstrasse where they were enraptured by the architecture and the decoration. As always, they did not omit to visit cemeteries, the old and new 'Gottessacker', where Piazzi marvelled at the innumerable monuments of marble, bronze and porcelain.

Piazzi Smyth's astronomical business was principally with the two firms of telescope-makers, Steinheil and Merz. Steinheil's firm, founded in 1854 by Carl August von Steinheil, was now operated by his two sons, neither of whom struck Piazzi Smyth 'as men of great genius or ability, with none of the vivacity and inventive faculty of the father as I saw him in Paris in 1855'. The father, a Professor at the Academy of Sciences, whom Piazzi wished to meet again, was unfortunately away from Munich. At Steinheil's Piazzi discussed prices for an eight-inch objective of focal length about ten feet which would fit into his dome in Edinburgh, and for smaller optics which he bought for his own use. He did not report seeing there any larger telescopes or glass-silvered reflectors though Steinheil had been one of the first opticians to introduce this technique of mirror-making several years earlier.

The work of the firm of George Merz, operated jointly by him and his two sons, Piazzi Smyth already knew from the large 15-inch refractor built for the Pulkova Observatory which he had been able to use there. He was also greatly interested in seeing the testing of the 13-inch refractor then almost complete for the Royal Observatory at Greenwich. Impressed though he was by the quality of their work he was prejudiced against the firm of Merz because of their reputed secretiveness about their art of lens-making which had been passed down from father to son for three generations. Because of this, he and W S Jacob, back in 1847, had chosen the Paris opticians of Lerebours and Secretan in preference to Merz for the six-inch telescope for Madras Observatory in India. At that time Piazzi Smyth had held up as an example of scientific openness the action of Daguerre who had immediately published his discovery of photography and allowed others to profit from it. His unease in Merz's workshop was not diminished when he observed a picture of the Madonna hanging between the busts of the illustrious founders of the firm, Utzschneider and Fraunhofer; according to him such a display was 'very much against one's ideas of Christianity and Biblical inspiration'. However, the brothers Merz were able to show him large equatorial telescopes under construction with excellent mountings which they could supply almost off the shelf and they were prepared to give him a definite quotation for a first-class equatorial telescope including mounting and dome of £7000 all told.

At the Bavarian Academy of Sciences Piazzi Smyth found considerable activity with the ruling of standard metre lengths in preparation for the introduction of the metric system into Bavaria at the beginning of the following year, 1870. The same activity was going on at the university, where Piazzi Smyth was received by the Professor of Physics, Philip Jolly, who showed him his laboratories full of electrical and optical apparatus but now dominated by a dividing machine ruling glass metres. Jolly was 'a desperate metre man' who described the metre as 'parfait, parfait!' and showed no sentiment whatever when Piazzi questioned him on the impending destruction of traditional German weights and measures.

By far the most interesting person whom Piazzi Smyth met in Munich was John Lamont, or Johann von Lamont as he was called there, Astronomer Royal of Bavaria and Professor of Astronomy in the University of Munich, famous for his researches on terrestrial magnetism. A native of Braemar in Scotland, Lamont had been sent to Bavaria as a boy to be educated in the Scottish Benedictine Monastery at Ratisbon on the Danube. Tutored in mathematics by the Prior of the monastery he gained many distinctions at school and at university with the result that at the early age of 22 he became a member of the Bavarian Academy of Sciences and at 30 Director of the Munich Observatory. In his native Scotland his distinction was recognised by an Honorary Fellowship of the Royal Society of Edinburgh. Lamont's investigations into terrestrial magnetism had led him to the discovery in 1845 of a 10.3 year periodic cycle of variation in the Earth's magnetic field which Rudolf Wolf, whose name is ever associated with sunspot numbers, showed could be described equally well by the sunspot cycle of 11.1 years. Lamont's discovery thus pointed to a correlation between the Earth's magnetic field and solar phenomena and at the time of Piazzi Smyth's visit he had just published his findings which associated aurorae with magnetic storms.

John Lamont, a pencil sketch in Piazzi Smyth's diary (RSE).

Piazzi Smyth knew very little about Lamont when he called at the observatory at Bogenhausen on the outskirts of Munich to make a courtesy call and present copies of the Edinburgh publications. Jessica and he had to walk some distance along country roads before they found a barely marked track which led to the observatory. 'What a road to a Royal Establishment!' The observatory turned out to be a small one-storey building and as the pair approached the entrance 'half a dozen rabbits disputed the way'. Lamont appeared, 'serenely smiling, a philosophical old man, say 70 years (he was actually 64) in a dressinggown of dark colour and a shirt perfectly innocent of collar or starch!' Inside the observatory Piazzi Smyth was astounded by the amazing variety of instruments, many of which had been designed and put together by Lamont himself. 'Never was an astronomer so independent of what the opticians chose to supply him; never was an observatory so full of all sorts of things that only the master could move about safely in it. Never were more stars observed by one man. Here is a man who for practical astronomy is practical above all others, not only in using instruments for observation but in designing them himself; and that, not by large funds, but by the help of young men selected for genius of the proper kind. He does not like to have *educated* astronomers as assistants.' Here was indeed a man after Piazzi Smyth's heart!

Lamont's standing in his visitor's eyes was even further enhanced when it became clear that they shared the same views on metrication. Lamont was a vehement opponent of the French metric system which he said was being imposed on Bavaria by the Prussians; being an ardent son of his adopted country he resented 'Prussian intrusion' in any form. Piazzi Smyth was delighted to be able to add Lamont's name to his party in his continuing anti-metre battle. On the scientific side he was very struck by Lamont's work in the field of geophysics which led him on his return home to examine the Edinburgh records of Earth temperatures and to search for periodicities and possible correlations with the sunspot cycle.

Some days later the Bavarians celebrated with great solemnity the Feast of Corpus Christi. From their hotel window the Smyths watched bemused as the colourful procession passed through the streets, in which Lamont took part as a high dignitary of the kingdom. Lamont tried to explain the significance of the Feast to his friends and that same evening Piazzi Smyth, his original prejudices melted away, recorded in his diary how, 'in the present age of freethinkers the people of Munich protest that they still firmly believe the fundamental truth of Christianity'. As on other occasions his extreme fundamentalist views tended to be somewhat theoretical and were apt to give way in personal encounters with people whom he liked. Carried away by his admiration for Lamont and by the enchantment of the beautiful and artistic city he left Munich half a Bavarian himself.

The Smyths travelled home via Paris where Piazzi wished to visit the French opticians for further information on telescopes. He kept no

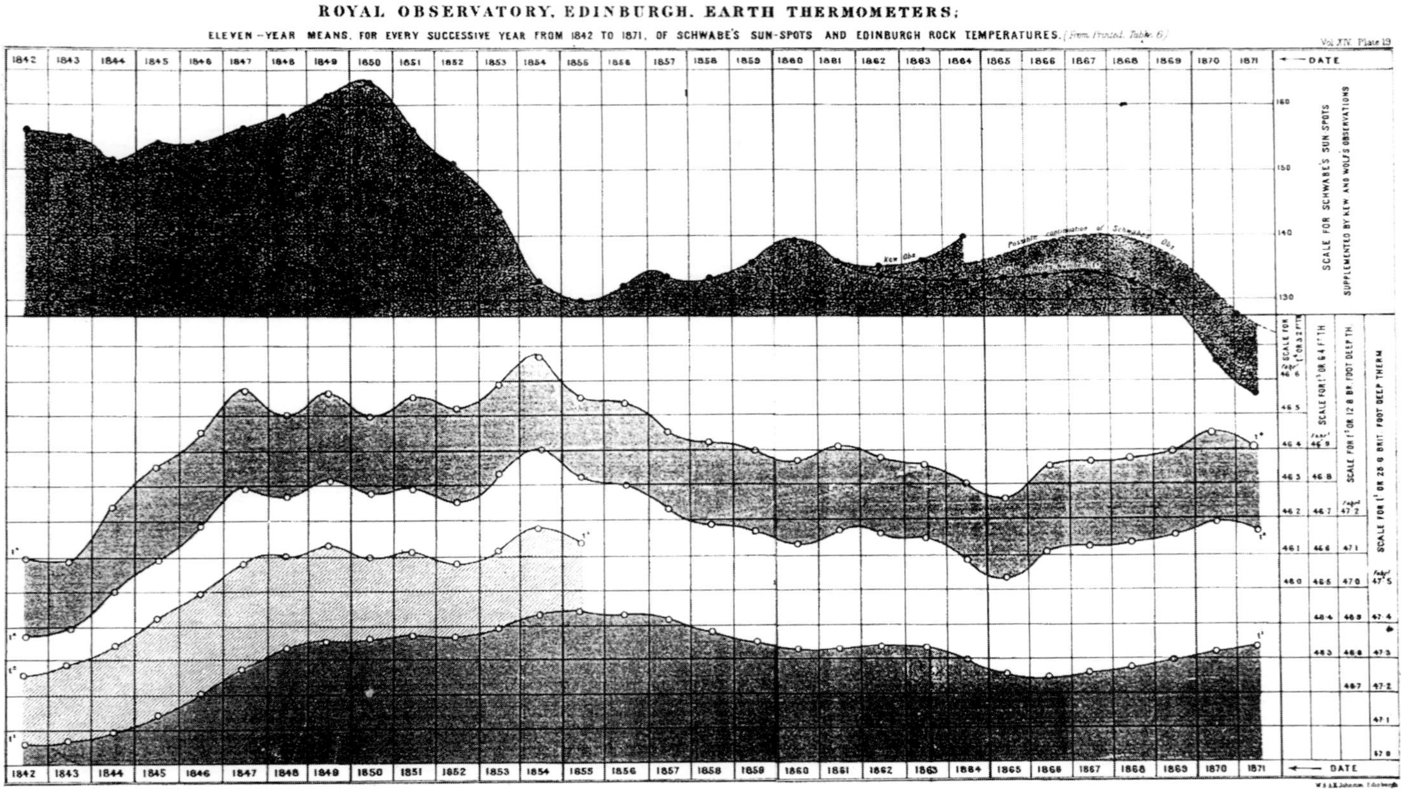

Cycles of sunspot numbers and Edinburgh earth thermometer readings (CPS, EAO vol. 14).

record of this visit; one of the opticians was Salleron, but it is most likely that his principal interest was in Leon Foucault and his laboratories which produced glass flat and parabolic mirrors. Foucault had already made 13-inch reflectors; his largest mirror was of 80 cm (31 inches) aperture and 45 metre focal length made from a disc supplied by the St Gobain glassworks.

Following his visits to the Munich and Paris opticians Piazzi Smyth could now put definite proposals to his Board. The price, £7000, quoted by Merz for a refractor and dome was excessive. An alternative to continental telescope-makers was the firm of Thomas Grubb of Dublin who in 1868 had completed a four-foot speculum reflector for the observatory of Melbourne in Australia, an instrument which at the time was much admired. The firm, which since the retirement of Thomas Grubb in 1868 was under the direction of his son Howard, had also made an equatorial telescope with an 18-inch Cassegrain reflector with metallic specula for Huggins in 1870. A serious constraint on the design and size of a telescope for the Calton Hill Observatory was the necessity to preserve the exterior of Playfair's building which limited the size of the instrument to the dimensions of the dome. A leading Scottish expert on the history of architecture, James Ferguson, was called in to advise on what degree of alteration could be allowed to accommodate a new equatorial. In the end Piazzi Smyth decided on a 24-inch reflector which would not only be considerably cheaper than the Merz 18-inch refractor but would be more than equal in power to it while still, with focal length of 120 inches, short enough to fit into the dome. Reflectors had become technically easier to construct as well as lighter following the invention in the late 1850s of the method of covering glass with a coat of silver. Large silvered-glass mirror telecopes were still, however, relatively rare and a 24-inch silvered mirror would be the largest of its kind yet made by the firm of Grubb. The new instrument was to have various attachments, in particular a large three-prism spectroscope and facilities for direct and spectroscopic photography intended to emulate the work of Huggins and others. A unique feature was that it should have 'a certain power of adaptation for being separated easily into component parts, transported with safety and re-erected with facility in an entirely different latitude'. Piazzi Smyth saw himself as 'preeminently the one above all British astronomers to be charged with the duty of cultivating "peripatetic" as well as "stationary" astronomy', and of repeating the Tenerife experiment with better equipment.

At a meeting of his Board of Visitors in July 1871 Piazzi Smyth explained these ideas to the members, and a week later Howard Grubb came across from Dublin to Edinburgh to open detailed discussions on the proposed new equatorial. As a result a formal contract was placed for a 24-inch equatorial telescope with the firm of Grubb by the Commissioners of H M Works and Public Buildings, who represented the Royal Observatory, and signed by their Secretary G Russell on 14 October 1871.

The terms of the contract were extraordinarily simple, saying essentially no more than that the telescope with subsidiary apparatus, a 15-foot revolving dome and observing chair, at a total cost of £2350, should 'work well to the satisfaction of the Astronomer Royal for Scotland'. An extra £400 was granted for laboratory apparatus, also loosely defined, such as 'chemicals for celestial photography' and 'Geissler tubes etc'. The instrument was to be completed within fourteen months, the delivery date being 14 December 1872. By the standards of the day this was to be a very substantial telescope of advanced design, one of the first large silver-on-glass mirrors to be made anywhere. An equatorial mounting was still considered a feat in the case of a large instrument. It is strange that the Board of Visitors allowed this clearly important and quite expensive contract to be placed without seeking expert advice. Grubb, for their part, offered a hostage to fortune in promising that the instrument would be to Piazzi Smyth's satisfaction without any qualification whatsoever. The mounting was not mentioned at all in the contract. The instrument turned out to be much more complex than Piazzi Smyth imagined; but for the moment he had no worries as he looked forward to having his dearest wish fulfilled.

A result of Piazzi Smyth's visit to Munich and of his discussions with Lamont was his renewed interest in a longstanding programme inherited by him from his predecessor, the recording of temperature below ground in an investigation initiated by J D Forbes with a grant from the British Association for the Advancement of Science in 1836. With the cooperation of Henderson, then newly appointed, a set of four thermometers had been buried at various depths down to 25 feet in the rock of Calton Hill for the purpose of discovering something about the thermal conductivity of the rock and the variation of temperature with time. Piazzi Smyth had carried on these observations following the procedures laid down by Forbes, and papers first by Forbes and later by Piazzi Smyth had appeared from time to time. While the original purpose of the experiment had been to investigate the thermal properties of the rock itself, Piazzi Smyth's new interest was in searching for variations which might be associated with solar variability. When in 1870 he assembled the results of several decades'

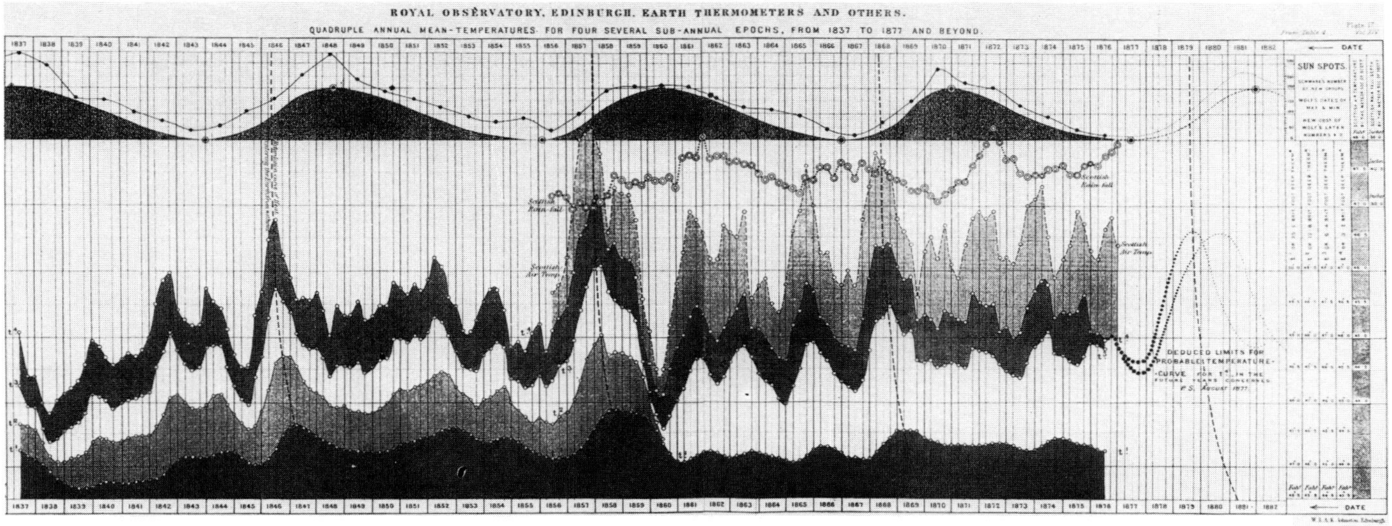

Analysis of earth thermometer readings and an attempt to correlate earth temperatures from 1837 onwards with sunspots and to extrapolate them into the future (CPS, EAO vol. 14).

earth temperatures in tabular and graphical form he thought he could recognise three separate cycles including one of 11.1 years, similar to the one which had been established by Lamont for geomagnetic phenomena and correlated with the sunspot cycle. He read a paper on his findings to the Royal Society of London an abstract of which was published in its *Proceedings*. One of the referees was not convinced by the 'cosmical' significance of the results.[1] Piazzi Smyth was furious, and protested at the 'misrepresentation on the part of a hidden referee'. He little realised that on this occasion the referee was an old friend of his, Balfour Stewart, formerly lecturer in Natural Philosophy at Edinburgh University and now Director of Kew Observatory. The paper was published in full in the *Edinburgh Astronomical Observations*. Tait considered it a 'remarkable' paper, which indeed it was. While realising that the apparent temperature variations were not directly correlated with sunspots—this followed from Langley's work on sunspots at the Allegheny Observatory in the United States which showed that the decrease in total solar radiation due to the spots would not be significant—he claimed an 11-year cycle correlated, he suggested, with solar prominences, an idea not altogether foolish in view of modern knowledge of variations in the corona and solar wind. He went so far as to predict, on the strength of the apparent 11-year periodicity, a heatwave to follow those of 1834, 1846, 1857 and 1868, for the year 1880. But before that date came a calamity befell the thermometers when a mad Portuguese sailor broke into the observatory and smashed them. Piazzi Smyth, never slow to impress on his Board the poverty-striken state of the observatory, reported that 'the

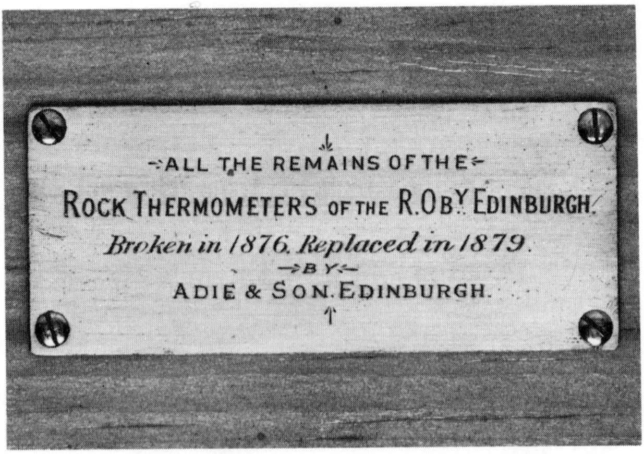

Inscribed plate on the box in which Piazzi Smyth preserved the relics of the original earth thermometers (RMS).

wretched man is now in a madhouse at public expense, while the observatory cannot afford to pay for replacements'! The thermometers were, in fact, replaced. Piazzi Smyth's attempt to relate solar and meteorological effects was one of many made during the 1870s which were not ultimately successful in establishing a simple connection between them. However, the Edinburgh temperature records, going back over a long period of time, remain of considerable value to this day, and the idea that there may be a correlation between weather and the sunspot cycle has found favour again in recent decades.

Chapter 9

Spectroscopy in Sicily

Spectroscopic studies in the 1850s saw many advances in laboratory investigations of the spectra of luminous gases for chemical analysis and for the discovery of new elements. However, as H Kayser has pointed out, the whole field of spectroscopy was lifted to a new level when in July 1859 G Kirchhoff published his researches on the relation between emission and absorption lines which were followed a year later by his joint work with R Bunsen on *Chemical Analysis by Spectrum Observations*. It then became possible to identify the presence of chemical elements in the Sun and any other luminous body in outer space, something which up till then had been generally thought to be impossible. Two years after the publication of his first fundamental paper Kirchhoff was already able to publish, in July 1861, his great map of the solar spectrum showing the presence in the Sun of a number of terrestrial elements including sodium whose everpresent 'yellow light' had masked the relation between spectral lines and chemical origin to John Herschel who, in Kayser's opinion, might otherwise have anticipated Kirchhoff by 30 or 40 years.

Kirchhoff and Bunsen's work was followed by a wave of papers of astrophysical interest. The first in the field of astronomical spectroscopy was G B Donati of Florence who in 1860 took up Fraunhofer's early observations on the differences between the spectra of different stars. Three years later L M Rutherfurd of New York City entered the field with some early attempts to classify stellar spectra, and in 1863 W Huggins and W A Miller in London started their well-known spectroscopic observations. In the same year Father Angelo Secchi in Rome began his survey of stellar spectra which led him to the conclusion that, though stars were numerous, their spectra fell into a limited number of well-defined groups, which allowed stars to be classified according to their spectral type. Huggins' interest was in performing detailed studies of individual stars by comparing the dark lines in their spectra with the bright lines produced in laboratory sources. He discovered that stars similar to the Sun were not unusual and that the common chemical elements traced in the solar spectrum by Kirchhoff were at least in part present in the stars. Huggins also made the important discovery that the spectra of certain of those objects labelled nebulae such as planetary nebulae and the Nebula in Orion consisted only of bright lines proving that they are incandescent gases and not unresolved distant clusters of stars.

The year 1868 was a particular landmark in the history of solar spectroscopy.[1] In August a total eclipse of the Sun occurred in India which came to be known as 'The Great Eclipse' because of the successes of the expeditions which were sent to observe it from France and England. The spectrum of solar prominences was observed for the very first time and found to be composed of luminous lines including those of helium, an element unknown on Earth and first discovered on this occasion. On the day following the eclipse P J C Janssen, later the founder and Director of the Meudon Observatory near Paris, succeeded in observing the prominences with his spectroscope, a feat which was achieved independently a few months later by Norman Lockyer in London. An attempt to see prominences outside a total solar eclipse had been one of Piazzi Smyth's unsuccessful experiments on Tenerife more than ten years earlier when his best efforts at reducing the brightness of the sky had been of no avail.

The year 1868 also saw the very first photographic record of the far violet solar spectrum which was obtained by L M Rutherfurd in New York using a spectrograph with a diffraction grating made by himself. Again in that same year a map of the solar spectrum was published by A J Ångström in Uppsala which remained the 'definitive' one until H A Rowland of the Johns Hopkins University in Baltimore in the United States produced his major photographic atlas of the solar spectrum in the 1890s. The great advance of Ångström's over Kirchoff's map of 1861 lay not only in the many additional spectrum lines which it revealed but also in the fact that it established a very accurate scale of wavelength, recorded in units of 10^{-10} metres, units which came to be called 'angstrom units' and which were soon universally adopted in spectroscopy. The same remarkable year of 1868 was also notable for the first successful determination by William Huggins of the velocity of a star in the line of sight through the Doppler shift of the hydrogen $H\beta$ line in the spectrum of Sirius.

The Indian eclipse stimulated much popular interest in solar spectroscopy which resulted in the manufacture of a variety of visual spectroscopes most of them of the direct-vision type which G B Amici had introduced which contained a train of five prisms, two of flint and three of crown glass. When Piazzi Smyth purchased one of these spectroscopes—the observatory had none of its own—his old enthusiasm returned and he resolved to take up again the question of the separation in the Sun's spectrum of lines produced in the Earth's atmosphere and those of genuine solar origin.

Surprisingly little progress had been made on this particular question since Piazzi Smyth's observations on Tenerife and there was still considerable disagreement as to the true source of the various absorption features. Piazzi Smyth himself felt sure that his observations of high and low sun on Tenerife had clearly demonstrated that the solar line spectrum consists of both terrestrial and solar components. Another who held the same view was Janssen whose interest in astronomical spectroscopy arose through his acquaintance with Secchi and who had set up his own observatory in Montmartre in 1862 for the express purpose of investigating atmospheric absorption in the solar spectrum. His goal was to see if there was not room for the premiss 'of a double origin for the rays discovered by Wollaston and Fraunhofer in the prismatic image of the Sun' between the extreme claims of Brewster on the one hand that all dark lines are telluric and of Kirchhoff who attributed them to the Sun's own atmosphere. He was unaware at that time of Piazzi Smyth's Tenerife work and of Kirchoff's surmise that certain spectral features might well be of terrestrial origin. In 1863 Janssen had made an important contribution to the question of the likely origin of the telluric lines. When studying the spectra of distant fires across Lake Geneva he saw absorption features which he recognised as existing also in the spectrum of the Sun. These appeared to him to be due to the effect of moisture and their presence in the solar spectrum to the effect of a moist atmosphere. Afterwards in Paris he investigated the actual absorption spectrum of water vapour by viewing bright lights through a 120-foot long tube filled with steam. While there seemed therefore to be evidence for the existence of water vapour in the solar spectrum, Ångström pointed out at about the same time that certain features in the solar spectrum, identified by him as telluric in origin by a comparison of high and low Sun, could not possibly be due to water vapour since he had observed them under very dry frosty conditions.

It had thus become accepted that the solar spectrum contained absorption lines of both solar and telluric origin, and that some of the latter were produced in dry, and others in moist air. The separation of lines into these different categories was a substantial task, with ever more lines being recorded with improved spectroscopes; and also an important task, as much for a study of the Earth's atmosphere as for the Sun's. Janssen embarked on the project in a systematic way and was awarded a grant from the French authorities to enable him to go to a clearer climate to study the changes in the Sun's spectrum near the horizon.

While keenly interested in the question of absorption lines produced by the Earth's atmosphere Piazzi Smyth was also drawn to the controversial problem of the spectra of the Aurora Borealis and of the Zodiacal Light. In the winter of 1867–8 Ångström observed from Sweden the spectra of several displays of the Aurora and one spectrum of what he believed to be the Zodiacal Light. He reported on

Normal Aurora, 1871, (top) and 'Aurora Superb' (bottom) of 25 October 1971 seen from Edinburgh, painted by CPS (EAO vol. 14)

these observations in the same famous year of 1868. His main result was the discovery that the auroral spectrum was a line spectrum, which always displayed one distinctive green line in emission, afterwards often referred to as 'Ångström's line'. He claimed that the same green line was also present in the Zodiacal Light, 'a circumstance', he wrote, 'of cosmical importance' pointing to a common origin for both phenomena. For the majority of astronomers the work of the great Ångström was deemed to have settled the matter—but not for Piazzi Smyth who was too familiar with the Zodiacal Light from his years in the clear skies of South Africa and from his time on Tenerife to believe that it had a line spectrum like that of the Aurora. He had described the appearance of the Zodiacal Light in some of his earliest papers and in his *Teneriffe Report*, and now in the northerly latitude of Scotland he was well placed to examine the Aurora. He was convinced that the two phenomena were entirely unconnected.

With characteristic energy he set about reading all the literature on both solar and laboratory spectroscopy which had been published in the ten years since his *Teneriffe Report*, filling his notebooks with reading lists and spectroscopic plans. His renewed interest in spectroscopy was considerably helped by his friendship with William Swan, one of the pioneers, known particularly for his identification of the 'Swan bands' due to carbon which appear in the spectra of the Sun and of comets. Swan, a man of Piazzi Smyth's own generation and a native of Edinburgh, who, having taught for some years in the Scottish Naval and Military Academy, held since 1859 the Chair of Natural Philosophy in the University of St Andrews. He was a very active Fellow of the Royal Society of Edinburgh and could give advice and assistance to Piazzi Smyth in his laboratory work.

Another very helpful friend and collaborator in this connection was Alexander Herschel, one of Sir John's sons who had been born in South Africa in the very year of Piazzi Smyth's arrival at the Cape and who had been known to him as a child. He had become a Lecturer in Physics in Glasgow University in 1866 and the first Professor of Physics at Durham College in Newcastle upon Tyne in 1871. Though Alexander Herschel is now mainly remembered for his contributions to meteor astronomy, he was also an accomplished spectroscopist who was able to set up a well-equipped spectroscopic laboratory in Newcastle. It was he more than anyone who encouraged Piazzi Smyth to take up pure laboratory spectroscopy in addition to his astronomical interest. Herschel, a bachelor, spent many a spell in the Smyths' house where he was a great favourite of Jessica.

Piazzi Smyth entered—or re-entered—the field of spectroscopy at a very favourable time. It was a new, exciting and rapidly expanding field in which there was plenty of scope for experimentation and for the design of apparatus. Advances in spectroscopy were still entirely experimental. There was as yet no theoretical explanation why certain elements exhibited certain spectrum lines or why one and the same element might show different lines under different conditions of temperature or why the spectrum of a molecule bore no resemblance to the spectra of its constituent atoms. How little was understood of the theoretical basis of spectroscopy at that time is clearly shown in the set of lectures which H E Roscoe, Professor of Chemistry in Manchester, gave on 'Spectrum Analysis' in 1868 and published as a book which was widely used as an introduction to the subject. Experimental spectroscopists were concerned with the design and construction of different types of spectroscopes, with the production of electric arcs and vacuum tubes for the excitation of spectra, and with the measurement of wavelengths and attempts at formal analysis of what were frequently quite complex sets of spectral lines. In all aspects of such work Piazzi Smyth made himself a master. With his love of experimentation and his deftness with apparatus he was to become as well known in the world of spectroscopy as he had been in that of mountain astronomy. Spectroscopy was then at a stage very similar to that in which astronomy had found itself at the beginning of the nineteenth century when William Herschel's famous motto was the guide '*quicquid nitet notandum*—whatever shines should be observed', meaning that it was only by accumulating observations that astronomers—and now spectroscopists—could expect to gain an understanding of the world.

The new field of spectroscopy brought a fresh purpose to astronomy. Up to the 1860s astronomers in publicly supported institutions like Greenwich, Edinburgh and their equivalents throughout the world had been primarily concerned with positional astronomy. At the same time, amateur astronomers in their private observatories had become innovators in the construction of telescopes for planetary studies, or, like Piazzi Smyth's father, found a worthwhile purpose in demonstrating Newton's laws of motion by observing the orbits of double stars, or by searching for new comets. As the exciting but more difficult field of spectroscopy opened up some of the old private observatories in Britain faded out while others, notably Huggins', adopted spectroscopy from the beginning. Many physicists in university laboratories also took up spectroscopic studies of the Sun. This general change from positional to physical astronomy coincided with the appearance on the astronomical stage of American astronomers and American observatories. When Piazzi Smyth took up office in 1846 his observatory mailing list consisted of 115 observatories and scientific institutions of which only 13, including just two observatories, at Washington and Georgetown, were on the other side of the Atlantic. Of 84 individuals who were deemed worthy of receiving personal copies of the Edinburgh publications only one was American—Professor Elias Loomis of New York University. By 1870 the position had changed radically. The United States—if Dreyer's list[2] of working observatories is accepted—now had no fewer than 16 active establishments most of which had been founded through the generosity of private donors and were attached to universities and colleges. Piazzi Smyth was well aware of the rapid progress of astronomy in the United States and, by sending them his publications, established very friendly contacts with some of the most interesting of the new generation of American scientists. In Europe, too, though unfortunately not in Great Britain, governments were persuaded to support the foundation of a new type of observatory for physical astronomy. The first among these were the Astrophysical Observatory at Potsdam in Germany founded in 1874 and the Meudon Observatory near Paris founded in the following year, both devoted entirely to astronomical spectroscopy, photography and physical observations and both highly successful.

Piazzi Smyth's revived interest in spectroscopy benefitted in a practical way from the move in 1871 of his official headquarters to a spacious house in Royal Terrace. Designed by Playfair and completed in accordance with Piazzi Smyth's special requirements it was one of the most magnificent of the houses in that splendid Edinburgh Terrace inhabited by wealthy ship-owners who could watch from the windows of their drawingrooms their vessels arriving in the port of Leith. In this beautiful house—which is now a hotel—there was space on its five floors for a large residence, the observatory library and business room and two laboratories on the top floor, one south-facing and intended as a solar laboratory and the other facing north towards an open sky providing an ideal vantage point for observations of the Northern Lights which during the years around the exceptionally high sunspot maximum of 1870 were particularly frequent.

When in August 1871 the British Association for the Advancement of Science met in Edinburgh Piazzi Smyth was too busy observing the almost nightly auroral displays to find time to contribute to the discussions. However, he attended the sessions and had the pleasure of hearing Sir William Thomson in the course of his Presidential Address praise the Tenerife work in his outline of advances in physics and astronomy. During the meeting the Smyths entertained the visiting astronomers in their new home when the guest list included many illustrious men such as John Couch Adams, James Clerk Maxwell, Lord Lindsay, David Gill, William Huggins, Jules Janssen, Franz Friedrich Brünnow and Abbé Moigno. Their local friend R M

Smith, Alexander Wallace from the observatory and Piazzi's brother Henry, by this time a distinguished figure in the Army, were also of the company.

Piazzi Smyth observed several displays of the Aurora during this period in all of which he found in the spectrum the green, or as he preferred to call it, the 'citron', line of Ångström. In the brightest displays he could also see the red line, discovered in 1870 by Zöllner in Germany, and even the four very faint blue and violet lines, a remarkable tribute to the power of his home-made spectroscope and to his own observational skill. He measured the wavelength of the citron line using for comparison a group of lines in the same green region produced by an acetylene lamp whose wavelengths had been supplied to him by Swan—he was rightly critical of the ubiquitous use of the yellow lines of sodium as wavelength standards. With his acetylene lamp he obtained a wavelength of 5579 angstrom units for the green line, very close to its true value (5577) and considerably more accurate than any previous estimate. This result, based on aurorae observed in 1870–1, was published both in *Monthly Notices of the Royal Astronomical Society* and in the *Edinburgh Astronomical Observations* in 1871. The German spectroscopist Vogel, then at the beginning of his distinguished career, independently published in the same year his observations of the spectra of some of the same aurorae, but without reaching the absolute accuracy of Piazzi Smyth's method of wavelength measurement.

The question of the precision of wavelength determination became particularly important after the discovery by E C Pickering of Harvard and C A Young of Princeton at the solar eclipse of 1869 of a green line in the spectrum of the solar corona which was believed by many to be caused by the same agent as the green line of the Aurora. Indeed there were spectroscopists who thought that the spectra of the Aurora, the Zodiacal Light and the Solar Corona all came from the same source, probably iron. There were so many lines in the iron spectrum that it was not difficult to find one close to the observed green ones and Piazzi Smyth was not alone in being doubtful about this particular identification. But he was also extremely sceptical about the first part of the proposition. To prove his point it was necessary to measure the wavelength of the auroral green line with the greatest possible precision and to scrutinise carefully the spectrum of the Zodiacal Light which, he believed, would display no emission line whatsoever.

In the spring of 1871, using a small direct-vision spectroscope made by the Edinburgh firm of James M Bryson, Piazzi Smyth started to study the spectrum of the Zodiacal Light. It was faint, but he thought he could confirm his expectation that there were no emission lines present. He also made one of the first ever attempts to record visually the intensity distribution in both the continuous spectrum of the Zodiacal Light and that of the twilight through a comparison with the spectrum of a laboratory source. However, the scope for observing the Zodiacal Light from Edinburgh was limited. Edinburgh's location was too far north and the transparency of its atmosphere too poor, particularly low down in the westerly direction where the Zodiacal Light lay. To study the phenomenon satisfactorily a journey to a more suitable environment in a low geographical latitude in spring time was indicated.

Piazzi Smyth chose Egypt for this enterprise and made preparations to go there in the following spring when he could combine his spectroscopic work with further measurements of the interior passages and chambers of the Great Pyramid. Having tried without success through the intervention of his Board of Visitors to obtain funds from the Treasury to carry out the pyramid work, he applied to the Royal Astronomical Society explaining that he proposed to travel to Egypt 'to observe the spectrum of the Zodiacal Light and to resume my researches of 1865 into the astronomy, geodesy and archaeology of the Great Pyramid' and requesting them to apply to the Treasury on his behalf for a sum of 'not less than £2000' to pay for labour at the pyramid 'which I have no means whatever out of my slender salary of hiring'. The council did not feel justified in approaching the government. He also applied to the Royal Society of London for a grant for a new Zodiacal Light spectroscope but this too was refused, ostensibly on the grounds that a spectroscope of the size he asked for would be unsuitable for such a faint source of illumination. Undeterred he fell back on his own resources and constructed a spectroscope of his own with an acetylene lamp for wavelength comparison. For the pyramid measurements he acquired a new set of rods divided in 'pyramid inches', photographic plates through his friend Sidebotham of Manchester, and a supply of magnesium for the flash photographs he proposed to take of the interior of the pyramid. All was ready for a second expedition to Egypt in the spring of 1872.

But now fate intervened. Just before Christmas Piazzi Smyth became seriously ill as a result of overwork and the effect of re-vaccination against smallpox. He was a whole month out of action, though he kept himself occupied reading books on ancient history. The hoped-for trip to Egypt, however, had to be cancelled and Piazzi Smyth was advised to take instead a long holiday in the Mediterranean. He and his wife, who was herself in poor health, were to spend three months abroad, including a month in Palermo in Sicily, a place with which Piazzi Smyth had sentimental ties as the home and observatory of his admired godfather. The trip was to be a rest cure but Piazzi Smyth did not neglect to pack his spectroscope and acetylene lamp as well as the meteorological kit of thermometer and aneroid barometer which accompanied him everywhere. This time sketchbook and paintbox replaced his camera.

Shortly before setting out Piazzi Smyth witnessed one of the most spectacular displays of the Northern Lights ever recorded and indeed the most completely observed Aurora for many years. It occurred on the night of 4 February 1872, and was seen by observers all over Britain and the continent of Europe. Reports came flooding in from latitudes as far south as Sicily, Egypt and India, and from many parts of the southern hemisphere. The Aurora was accompanied by magnetic disturbances which were recorded at Greenwich and also by such well-known observers as Secchi in Rome and Lamont in Munich. In an excellent account in *Nature* Piazzi Smyth described his own observations of the auroral spectrum during this famous display and his measurements of the wavelengths of the five known lines. These wavelengths, calibrated against those of his acetylene reference light source, were the most precise at the time. The Great Aurora was also observed spectroscopically in Rome by Father Secchi at the Collegio Romano and by Respighi at the Campidoglio Observatory. Respighi whose observations were published in *Nature* a few weeks after Piazzi Smyth's claimed that he had been able to observe from Rome the spectra of both the Aurora and the Zodiacal Light, 'demonstrating the identity of the Zodiacal Light and that of the Aurora'. In this he supported Ångström's view, but Piazzi Smyth was certain that both observers were mistaken. He made plans to repeat in Sicily the same spectroscopic observations of the Zodiacal Light with the same equipment which he had used on the Aurora at home, expecting to refute what he knew was a confusion on the part of Ångström and his Italian colleague.

From the moment the Smyths left home at the end of February 1872 until they returned in May, Piazzi's diaries bubble over with enthusiasm. The oppressive load of pyramid correspondence was cut off, and he became once more the witty intelligent observer of the world who had delighted readers of his Tenerife, Russian and Egyptian travel writings. He kept detailed notes, intending perhaps to publish a book on his journey as he had done on those earlier occasions; but it never materialised, not surprisingly in view of the endless demands of his importunate pyramid followers who robbed him of much of his spare time for the rest of his life.

The Smyths sailed from Liverpool on the SS *Kedar*, a steamship carrying 500 passengers. As reading matter for the voyage Piazzi Smyth bought Clerk Maxwell's newly published book on *Heat*, Griffin's *Algebra* and some religious literature. On board ship he took his usual intense interest in everything—the machinery of the ship itself about which he discoursed with the engineers, the lives of his fellow passengers, the state of the wind and weather. He took twice-daily meteorological readings on his instruments and recorded the colours of the sea and sky in numerous sketches.

Once the ship had passed Gibraltar the Zodiacal Light became brilliantly visible each evening for two hours after sunset, twenty times brighter in Piazzi Smyth's estimate than at Edinburgh: 'Ångström could never have seen it from Uppsala', he noted in his diary. In the Mediterranean, a day before arriving in Sicily, a storm blew up carrying fierce hot winds from the south west. Next morning in Palermo Bay the Sun rose blue-white in colour and the twilight sky on that evening was of a distinctly green hue. These unusual sights Piazzi Smyth painted on the spot; the pictures were reproduced later in the *Edinburgh Astronomical Observations*. He surmised that the blue

Zodiacal Light at Palermo, painted by CPS (EAO vol. 14).

sun and the green twilight were caused by the dust from the Sahara desert blown in on the previous day's storm. He was quite correct in this interpretation; and in fact his colleagues at Palermo could provide him with a sample of the dust collected from the roof of the observatory on the morning following the storm. Piazzi Smyth appears to have been the first to record a blue Sun in a scientific publication and to recognise its cause though without being able to offer a physical explanation of the phenomenon.[3]

Twelve days after leaving Liverpool the *Kedar* docked at Palermo. The Smyths were met at the ship by Colonel Henry Yule, a friend of long standing and now a resident of Palermo, 'looking like Marco Polo himself', a reference to Yule's well-known translation of the voyages of the Venetian traveller which had just been published. Henry Yule was a native of Dirleton, near Edinburgh, and had been a contemporary of Piazzi's brother Warington at Cambridge where he took the classes of the astronomer Challis. Apart from a brief spell teaching at the Scottish Royal Military Academy in Edinburgh he had spent most of his life in India, in various scientific and administrative capacities. He had become an authority on the geography and history of India and Asia, was the contributor of several scholarly articles on these subjects to the *Encyclopaedia Britannica* and was later knighted. When the Smyths visited him he had retired from the Army and was living in Palermo with his wife and only daughter. Yule was one of those who had warned Piazzi Smyth in the early stages of his pyramid activities of the cranks who were attaching themselves to him—a warning which unfortunately went unheeded. The hospitable Yules introduced their friends to the English colony in Sicily and showed them the sights of the town and the surrounding countryside. The astronomers at the observatory and the scientific professors at the university were no less welcoming. The month in Palermo was in every way a happy and successful time for the Smyths.

The Director of the Palermo Observatory at the time of the Smyth's visit was Gaetano Cacciatore who with his assistants M Tacchini and A Ricco belonged to the very active group of Italian spectroscopists of whom Father Angelo Secchi was the leading figure. Following the discovery of 1868 of the method of observing prominences at the sun's limb outside solar eclipses the Italian school had become especially active in the field of solar physics, using slit spectroscopes attached to their equatorial telescopes for prominence observation. In 1871 the Italian astronomers had founded the Society of Italian Spectroscopists which published *Memorie della Società degli Spettroscopisti*, a leading journal of astronomical spectroscopy edited by Tacchini which began publication in that very year.

Arriving at the Palermo Observatory the Smyths found a fine building and good instruments such as a new meridian circle from Pistor and Martins and a 9.5-inch Merz equatorial with a 20-prism prominence spectroscope attached to it. Piazzi Smyth had never seen prominences before and was shown how to trace them in the light of the red Hα line of hydrogen. He was delighted to see for the first time the yellow D3 line of helium, the solar element then unknown on earth, and to satisfy himself about its position relative to the two D lines of sodium by simultaneously viewing the spectra of a prominence and of a sodium flame. On an evening visit Piazzi and Jessica were shown Jupiter through the equatorial, and, though its image was not as good as they had seen it on Tenerife, they managed at least to recognise the planet's red spot.

Observations of the Zodiacal Light could only be made on moonless nights and while waiting for the right time to come round there were preparations to be made with the help of the Palermo astronomers who were convinced of the value of Piazzi Smyth's proposed observations. His colleagues were not familiar with the use of a comparison spectrum, so he instructed them by day in the operation of his acetylene lamp and prepared a list of comparison wavelengths. The crucial observation which was to prove that the Zodiacal Light emitted a continuous and not a line spectrum was to watch its spectrum while the broad slit of the spectroscope was gradually narrowed. A line spectrum should then become bright and clear while a continuous spectrum would merely become dimmer. Recording the result of the experiment Piazzi Smyth wrote 'Cacciatore also finds that on continually decreasing slit width until ZL spectrum vanishes, said spectrum retains to the last its large sensible breadth, i.e., it is a continuous spectrum without line or lines throughout the breadth it occupies.' Tacchini also agreed with this conclusion though it was somewhat embarrassing for him, as he had just published a note on his recent observation of the coronal spectrum at the Sicilian eclipse of 1870, stating as his considered opinion that the Aurora, Zodiacal Light and corona all had the same spectrum. 'But now I show him', wrote Piazzi Smyth, 'that the ZL spectrum is like any old residual twilight spectrum and also like the spectrum of the starlit sky or starshine. And why not, if the ZL is the united glare of millions of meteors shining by reflected sunlight, as I have always maintained?' To give a final touch to these conclusions, Jessica was appealed to, and joined in the observations on the last evening: 'Jessie declares that the ZL spectrum is as different from the auroral spectrum as night is from day.'

Later when sailing through the Mediterranean Piazzi Smyth also recorded the distribution of intensity within the continuous spectrum of the Zodiacal Light and found it similar in distribution to that which he had observed in the twilight sky and in sunlight reflected from a wall, a firm indication that its origin was indeed solar. The results of these important Palermo observations of the Zodiacal Light and of the Edinburgh observations of the Aurora were published in the *Monthly Notices of the Royal Astronomical Society* in 1872 in a paper which Richard Proctor, Secretary of the Society, described as 'masterly'. 'I have derived more satisfaction from its perusal than from any scientific paper I have read for many a month', he wrote. Piazzi Smyth was in a unique position to give a final verdict, having in the same season observed both phenomena each from a suitable latitude. Yet, though he had effectively solved the problem, his results did not immediately receive universal recognition. There was that undefinable bias which scientists meet in every generation towards interpretations given by well-established men. As Piazzi Smyth himself said later, 'everything from M Ångström deserves exceeding respect', and though Ångström was dead, his view that the Aurora and the Zodiacal Light were one and the same persisted. Vogel, observing also in 1872, believed that he could confirm the presence of the main auroral line in the spectrum of the Zodiacal Light; so did Lord Lindsay. Roscoe in the third edition of his *Spectrum Analysis* of 1873 describes the so-called 1474 line (Kirchhoff's label of the green line at 5577 angstroms) as one 'which links together the apparently unconnected phenomena of the solar corona, the Zodiacal Light and Aurora'. However, Scheiner's *Treatise on Astronomical Spectroscopy* (1890), the first to be exclusively devoted to the subject, gave Piazzi Smyth and his three Italian colleagues full credit and priority for establishing the perfectly continuous nature of the spectrum of the Zodiacal Light. During the following two years A W Wright working at the Yale Observatory in the United States not only confirmed Piazzi Smyth's view that the spectrum of the Zodiacal Light is continuous but also showed that its light is polarised, proving beyond doubt that it is reflected sunlight. The green auroral and coronal lines long remained mysteries; the auroral line (due to neutral atomic oxygen) was not identified until 1925 and the coronal line, attributed to an unknown element 'coronium', defied spectroscopists until 1942 when it was theoretically shown to be due to highly ionised iron.

The sunspot maximum of 1870 was an exceptionally high one and observing the Aurora became particularly popular after the many marvellous displays which accompanied it. To amateurs it became a great attraction. A delightful illustrated volume on the Aurora was published in 1879 by J Rand Capron, one of the most assiduous of auroral observers who contributed many valuable observations from his private observatory in Guildford in the south of England. This was 'respectfully dedicated to Professor Charles Piazzi Smyth, Astronomer Royal for Scotland, one of the earliest observers of the Aurora and Zodiacal Light'.

Jessica's role as an active astronomer was a surprise to the Sicilians. One evening when she felt too tired to accompany her husband to a social gathering in the Yules' home, the following conversation was gleefully recorded by Piazzi in his diary:

> Says Senator Professor Cannizzone, 'What is the matter with Mrs Piazzi Smyth?'
> Col. Yule: 'She has over fatigued herself and brought on a slight fever; she was actually twice at the Observatory on the one day'.
> Senator P.C. (with great astonishment): 'What has a donna to do at an Observatory!'

In between their astronomical activities Piazzi and Jessica visited the physics laboratory and, for Jessica's interest, the geological department of the university. They toured botanic gardens, the museums and art galleries—Piazzi as always noting down everything of interest and illustrating his descriptions with vivid watercolours. During this visit to Sicily, Piazzi mentions for the first time Jessica's slight lameness, a forewarning of the crippling disease from which she later suffered cruelly. The cruise had been intended for the benefit of Piazzi's health, but it was his wife who found it difficult to keep up with his unflagging energy. However, they were not deterred from setting off one morning at 7AM, mounted on donkeys, for Monte Pellegrino, which Piazzi was determined to see under the early morning lighting. As a memento of Palermo Piazzi purchased a beautiful *pietra dura* table which remained a treasure in the Smyths' home until his death 30 years later.

The next stage of their cruise took the Smyths along the Sicilian coast and past Messina Bay, giving splendid views of Mount Etna; then through the Ionian islands and northwards into the Adriatic Sea to Ancona. The next stop was at Trieste, then in Austria, where they visited the Nautical Academy which operated a hand-fired time-gun and where Piazzi Smyth explained to the Austrian director, Dr Lishman, his special method of electric firing.

At Venice their ship, the *Marathon*, anchored within a stone's throw of the Doge's Palace, a convenient base from which to explore the many attractions of that beautiful city. Ascending the Campanile of San Marco Piazzi spent two sessions producing some exceptionally pretty watercolours of the ship and the Church of San Giorgio

Maggiore. From Venice there was time to visit Padua by rail, where the Smyths were warmly received at the observatory and conducted round by G Lorenzoni, on behalf of the Director, G Santini, who was by then 90 years of age! Like the one in Palermo, the observatory in Padua was well equipped with a meridian circle, an equatorial telescope and the inevitable spectroscope for viewing solar prominences which was then all the fashion in Italy. Piazzi Smyth was undoubtedly deeply conscious of the shortcomings of his own run-down establishment.

Before leaving Venice the travellers paid a visit to the Frari Church which impressed itself deeply on Piazzi Smyth's mind. The Church of Santa Maria Gloriosa de Frari contains the tombs of many famous artists, among them the pyramid-shaped monument to Canova based on a design which Canova had originally intended for the tomb of Titian. It depicts figures in grief entering the tomb door of the pyramid; to Piazzi Smyth it was a work of art 'excelling in sublimity, simplicity and appropriateness'.

Throughout the 17-day return sea voyage the indefatigable Piazzi did not relax for a moment. He concentrated on observing the light of the ship's wake, a phenomenon he had already noticed on the outward journey especially in the Bay of Biscay and again in the Adriatic. He remarked that though the ship produced very little froth in the water by day, yet by night it had a long luminous trail full of glittering lights and 'sparkles'. Sea luminescence was well known to mariners and travellers, and by the middle of the nineteenth century had been recognised as caused by small organisms or 'animalcules' now known as dinoflagellates.[4] The sailors told Piazzi Smyth of 'discs and sparkles, splendid like gold' which they had seen near Cape Verde in Western Africa. With the help of one of the sailors Piazzi hauled up buckets of water from the phosphorescent wake, in which he was able to induce momentary sparkles by agitating the water; one night the deck was actually covered by a sheet of light when a wave broke over it. In order to capture the spectrum of the luminous sea Piazzi sat patiently through the night in a deck-chair with his instruments behind a black cloth to reduce stray light. In a series of experiments he compared the light of the sea with that of the starlit sky and ruled out the possibility that the phenomenon could be caused by reflection of any sort. Using his spectroscope with the comparison lamp he made a photometric record of the phosphorescent wake and found a continuous spectrum displaced towards the blue with respect to that of the Zodiacal Light. He placed the peak of the continuous spectrum at a wavelength of 5050 angstrom units. He also succeeded in estimating the brightness of the wake in terms of the Zodiacal Light. Piazzi Smyth's appears to have been the first ever

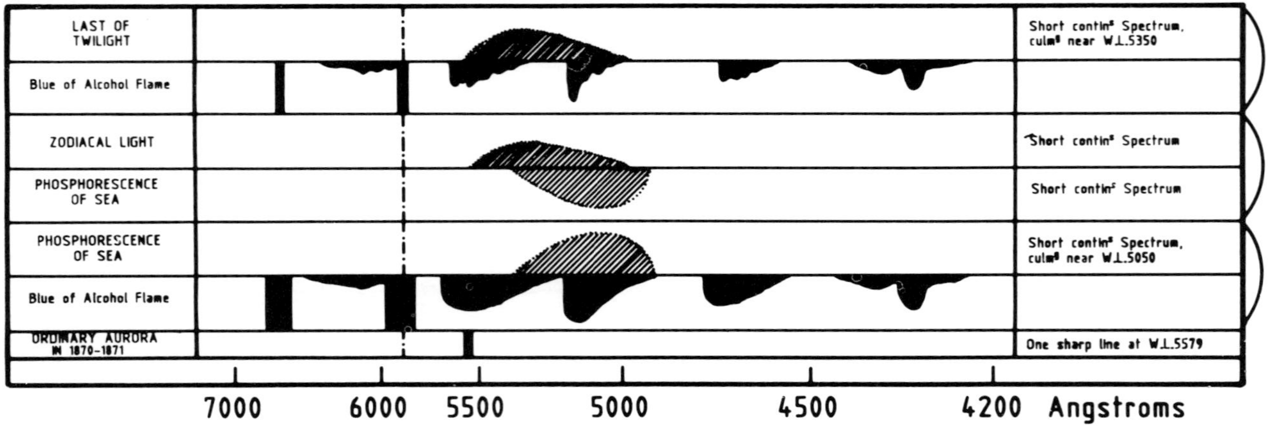

Comparison of the distribution of light in the Zodiacal Light and sea phosphorescence. Pairs of spectra were observed simultaneously: twilight and calibrating lamp; Zodiacal Light and sea phosphorescence; phosphorescence and calibrating lamp with wide spectroscope slit. The single-lined normal auroral spectrum is also shown (from CPS' diagram in MNRAS vol. 32).

spectroscopic observation of sea luminescence. His result, that the spectrum is continuous and displaced to the blue of the solar spectrum, was correct and the position he found of the peak light intensity is indeed remarkably close to the true value of 4750 angstroms. The observations were briefly described in Piazzi Smyth's paper on the Zodiacal Light in Palermo, published in *Monthly Notices of the Royal Astronomical Society*, though astronomers paid little heed to this unusual piece of research.

The details of Piazzi Smyth's spectroscopic work on the Aurora, the Zodiacal Light and sea luminescence, and of his observations of the blue sun and twilight phenomena, accompanied by beautiful coloured plates, appeared only five years later, in 1877, in volume 14 of the *Edinburgh Astronomical Observations*. These publications, which carried mainly ongoing positional measurements from the mural circle, each volume covering many years of observation, were a far from suitable vehicle for research papers, and interesting original results such as these went frequently unnoticed or appeared too late to make their full impact.

Chapter 10

Tribulations at Home

The Smyths had hardly returned from their Mediterranean cruise than they were on their travels again, this time to Dublin to inspect the state of work on the 24-inch equatorial reflector under construction in Grubb's workshop. By the summer of 1872 Grubb had made rapid progress and, in Edinburgh, the small dome on the top of the observatory building was being adapted for the reception of the new telescope which was to be installed before the end of the year. During the ten days he spent in Dublin Piazzi Smyth paid a number of visits to Grubb's establishment in Rathmines where he discussed the proposed attachments to the telescope for astronomical photography and spectroscopy. The work appeared to be progressing well and apart from some unease regarding the stability of the adjustable mounting Piazzi Smyth declared himself satisfied with the state of the instrument. He took time off to visit the Bank of Ireland—where Thomas Grubb had been an engineer before he founded the optical works—where he was shown the blocks used for printing bank-notes. He and his wife could also enjoy the company of their host, Wentworth Erck, a man of independent means who had a fine private observatory and laboratory near Bray in County Wicklow and had a certain standing as an observer of sunspots.

The Smyths visited the observatory of Trinity College, Dublin, situated at Dunsink ten miles outside the city, which Piazzi found to be considerably better equipped than his own observatory in Edinburgh. He was particularly impressed by the 12-inch refractor which had been mounted by Thomas Grubb in 1868 with an object glass by Cauchoix of Paris which had been donated by Sir James South to Trinity College, Dublin in 1863. Piazzi Smyth admired Grubb's workmanship which boded well for the quality of his own telescope.

The Director of Dunsink Observatory and Astronomer Royal for

Ireland at the time was F Brünnow who had moved to Ireland from Ann Arbor in the United States and who was widely known for his work in the field of spherical astronomy—the ideal person to check Piazzi Smyth's calculations on the position of the celestial pole at the time of the conjunction of the Pleiades with the equinox, his favoured date for the building of the Great Pyramid of Gizeh. Brünnow's figures were incorporated with great flourish in the later editions of *Our Inheritance in the Great Pyramid*. Piazzi Smyth's enjoyment of the hospitable reception at Dunsink was complete when he found Brünnow's young son and his father-in-law, formerly a Presbyterian minister and President of Michigan College, 'well pleased to listen to Pyramid talk'.

Piazzi Smyth returned home in happy anticipation of the arrival in Edinburgh of his longed-for 24-inch reflector. Before leaving Ireland he had asked David Gill as a trusted friend to make an independent inspection of the telescope while he was in Dublin to see the 15-inch refractor which Lord Lindsay had ordered from Grubb for his observatory at Dunecht. Gill reported to Piazzi Smyth: 'If a fault could be found I should have said the lower part of the mounting was lighter than I should have liked, but Mr Grubb explained to me that there were reasons for that, viz., that you wished the stand to be transportable. I of course keep this quiet, as you wished me to do.' Piazzi Smyth, obviously, did not wish to spread abroad his idea of using his new telescope as a movable instrument for a 'peripatetic observatory'. Gill also commented on the limited space which would be available in the Edinburgh dome for an instrument of this size. However, Piazzi Smyth's confidence in the competence of Grubb as engineers was not seriously shaken by Gill's comments and he no doubt endorsed the description of the telescope which *Nature* published in September 1872 when its construction was well advanced:

> We are glad to see that Professor Piazzi Smyth, Astronomer Royal for Scotland, has at last got the reward of his twenty years' persevering and creditable importunity in the shape of a new equatorial telescope for the Royal Observatory on the Calton Hill, Edinburgh, for which the Government last year granted £2300. Hitherto, as the "Scotsman" justly remarks, the Metropolitan Observatory of Scotland has been in the position of the meanest appointed Government astronomical institution in Scotland. The new telescope will have an aperture of two feet upon a focal length of only ten feet, a larger diameter in proportion to focal length than any astronomical telescope yet introduced into any observatory; and it will no doubt be by far the most powerful instrument ever erected under so small a dome. The instrument, which is being built by Mr Howard Grubb of Dublin, is to be constructed on

> M. Foucault's comparatively new principle of having the speculum of glass coated with pure silver. The instrument is to be employed both for photography and spectroscopy. Both these classes of research require the seemingly impossible accompaniments—that the telescope must have the utmost amount of firmness, and also have the most accurate possible movement at the same rate at which the stars change their position in the sky. Notable features in the new telescope will therefore be the remarkably perfect clockwork apparatus, the several devices connected with the prisms of the spectroscope, the means by which not merely celestial objects will be kept in view, but those by which the rays of chemical flames will be brought into comparison with the light of the stars. December next is the time fixed for the completion of the new instrument, but meanwhile active preparations are being made in the Royal Observatory for its reception.

The telescope arrived in Edinburgh as promised in December 1872 but on account of unusually bad weather no tests of its performance could be made before the formal deadline of 14 December. David Gill in his Christmas letter wrote feelingly: 'I have been thinking of you, your telescope and the weather—it has been miserable.' In spite of that, and in spite of the fact that the alterations to the dome were not finished, Piazzi Smyth confidently authorised the payment of their account to Grubb in a glowing letter to the Ministry of Works. Indeed, he was carried away by his enthusiasm: 'I have now within a week of the closing time stipulated by Mr Grubb's contract the pleasure of reporting that in all its main essentials his work is not only performed but admirably performed; so that the large Edinburgh Equatorial so long looked forward to by the scientific public here, is now at last a reality, is mounted in its place and is nearly ready to begin a useful course of observation.'

But no sooner had Howard Grubb and his engineers left on Christmas Eve with the intention of returning shortly to finish some details of their work than Piazzi Smyth realised that the instrument was by no means in proper functioning order. He told the Ministry; 'I was perhaps premature in proposing that the whole of the balance of the contract money should be paid.' For his part Grubb was amazed to receive a long list of complaints a couple of weeks later. Grubb's engineers on their return to Edinburgh in the New Year suggested that the trouble had arisen because Piazzi Smyth had tried to use the instrument before it was completed. An acrimonious correspondence followed in which Piazzi Smyth demanded more and more changes, insisting that according to the contract the instrument had to 'satisfy the Astronomer Royal for Scotland'. Howard Grubb tried to be patient but could not be expected to accede to a list of 26 alterations which Piazzi Smyth drew up in February 1873.

There is little doubt that Piazzi Smyth had to a large extent himself to blame for the ensuing trouble. The contract with Grubb was far too loose; yet if he had only allowed the firm to complete the telescope in the way they wanted, a working instrument might well have been installed, even if, as it afterwards appeared, the figure of the mirror was not perfect. In that case it would have at least been possible to give an independent judgement on Grubb's work. Piazzi Smyth had also failed to appreciate how very tightly the telescope fitted the small dome until he actually saw it in its place. This was no fault of the contractor but was the result of Piazzi Smyth's insistence on having the largest possible telescope which could be squeezed into the available space. It was also unfair of Piazzi Smyth to criticise Grubb's spectroscope with hindsight after he had become familiar with spectroscope design through his dealings with the Paris opticians.

One of Piazzi Smyth's first complaints had been that the silver surface on the mirror was so poor from the beginning as to prohibit making optical tests. He pointed out that Grubb had no experience in making silvered mirrors, and that indeed this was the first one that they had made. When after two years the silver coat on the primary mirror had deteriorated badly the Office of Works suggested that the mirror ought to be tested without its silver coat. This was an opportunity to call in as an outside expert Henry Cooper Key, a respected amateur astronomer and successful telescope maker who specialised in silvered mirrors. It was he who had supplied David Gill with his very first telescope. Cooper Key's verdict was that the figure of the primary mirror was 'positively bad' and unfit for its intended purposes and that the secondary mirror was too small. Cooper Key suggested that the mirror ought to be refigured, and the Board of Visitors in 1875 supported Piazzi Smyth in calling for Grubb either to replace the mirror or to refigure it. Grubb called for the appointment of an independent arbiter, and after some discussion W M H Christie, then Chief Assistant at Greenwich, agreed to serve in this capacity. Christie's opinion was that the telescope was entirely fit for use, and that it needed only a larger secondary mirror—which Grubb readily agreed to. The Board dropped its demand for the refiguring of the mirror and paid Grubb the balance of their account. However, Piazzi Smyth stubbornly refused to accept Christie's verdict and continued to assert that the contract had not been completed to his satisfaction. 'I should remind you that I have not given up either to Mr Christie or to anyone living my duty according to the contract of judging to the best of my ability as to when the apparatus has been supplied in good order and works well for its intended astronomical purposes.' He drew up a list of extras and improvements which would produce a workable telescope but when he asked for a special grant of £1400, the money, not surprisingly, was refused.

There have frequently been disagreements between scientists and the makers of their instruments, but the story of the Edinburgh equatorial was a major tragedy. Piazzi Smyth had waited 26 years for his great peripatetic telescope only to find a largely useless instrument in the observatory for his remaining 16 years in office. The disappointment from which he never quite recovered was all the greater as the telescope had promised to be a unique instrument of advanced design with a silver-on-glass mirror larger than any other in the United Kingdom. Apart from David Gill nobody had questioned the particular feature of its movability which was to be one of the reasons for its ultimate failure. Howard Grubb and his engineers had also obviously underestimated the complexity of their task.

The troubles with the equatorial were not the only ones for Piazzi Smyth at this time. When after his return from Munich he tested the performance of the transit circle, now more than 30 years old, he discovered serious faults in its piers. A review of the observations made with the instrument since it first came into use brought some disquieting facts to light. There were fluctuations in both azimuth and level of the instrument which correlated with the temperature of the soil as recorded by the shallowest of the rock thermometers. There was also a long-term change in these errors over the years, confirming a suspicion which he had in his early days that the stone piers were sensitive to very small fluctuations in temperature, even to those caused by the illuminating hand lamps in the dome. The irony of the situation was that the piers had been made not of the native underlying stone of the Calton Hill but from sandstone specially brought in from the Craigleith quarry, a type of stone which, according to Piazzi Smyth, was particularly unsuitable. It was a blow to him to realise that the fruits of long years of toil, since Henderson's first day at the observatory, were to be worth less than he had formerly claimed. The results were still of some value, but Piazzi Smyth decided to terminate the observations—apart of course from the time service—and to concentrate on completing the reductions and on publishing the whole series of observations covering a period of 20 years in the last volumes of *Edinburgh Astronomical Observations*. In any case the sky in Edinburgh, from a combination of smoke from the greatly increased number of factory chimneys and the bright gaslights, blotted out the faint stars which were an essential ingredient in Piazzi Smyth's original 'solar orbit' project. The decision to abandon the astrometic work on account of deteriorating sky conditions on Calton Hill proved, incidentally, that Piazzi Smyth had not intended the home observatory as the permanent site of the equatorial.

Instrumental problems apart, there was an ever-present lack of general funds. Money was needed to print the *Edinburgh Astronomical Observations* and to bind the books in the library. The Board of Visitors faithfully supported Piazzi Smyth in his annual requests for financial support and in 1871 had set up a special committee to press the case for a 'proper Royal Observatory'. However, in spite of their pleas, even a contribution towards immediate essential expenses was turned down. 'The Lords Commissioners decline to sanction any increase of the grant of money for the expenses of the Royal Observatory of Edinburgh' was the curt reply received in July 1872. David Gill wrote sympathetically to Piazzi Smyth:

> I am exceedingly sorry to hear that Government has all at once turned round upon you with so niggard and snappish a reply. Such an answer however I would not feel disposed to look upon in its literal sense. A Chancellor of the Exchequer looks on the matter in quite another light than does our Scottish Astronomer Royal. He does not see nor can he understand the necessity or application of telescopic adjuncts. He looks up his books and finds: "Oh—that observatory and its astronomer have got a new big house and a new big telescope, what more on earth can they want—write and say he'll have no more money—and there's an end of it. I think a year's rest and possibly a change in Government and another application with pressure will make it all right.

In 1873 things changed at last. It was the year when Sir William Thomson became President of the Royal Society of Edinburgh and thus an *ex officio* member of the Board of Visitors. Piazzi Smyth was impatient for Tait to get his influential friend to work without delay on the observatory's problem, but Tait reminded him: 'You forget your Physics. The mass of a President RSE is great, and you must at one time have known that finite force, such as I can apply, does not rapidly get a great mass under weigh—nor, when it has gone sufficiently far, does it easily stop again.' But Thomson did indeed take up the cause of the observatory. With the past President of the Royal Society of Edinburgh, Christison, he prepared a strong letter to the Secretary of State at the Home Office which recapitulated the history of the observatory's transfer from the Astronomical Institution to government ownership in 1847 and the responsibility which the Government had taken on to maintain it. The letter also drew attention to the unsatisfactory working conditions and low level of salary of Piazzi Smyth and his assistants.

Sir Lyon Playfair, Member of Parliament for the Universities of Edinburgh and St Andrews, drew the Secretary of State's attention to this letter, and sent copies to John Couch Adams and Warren de la Rue, past and present Presidents of the Royal Astronomical Society. Both of these eminent astronomers supported Thomson and Christison's letter and brought the matter before the Council of the Society, De la Rue wrote: 'I think that the Royal Observatory of Edinburgh is undermanned, the staff underpaid, and that if it is decided to continue the Observatory, the proposed increase is not too great. There is much work to be done besides that which comes within the scope of the Greenwich Observatory, and although the site of the Edinburgh Observatory is not of the best, yet it could contribute greatly to the advancement of Astronomical Science.' Adams' response was: 'My own opinion is strongly in favour of maintaining the Edinburgh Royal Observatory; certainly it is very shabbily treated at present.'

The intervention of such an influential scientist as Thomson and the support he mobilised could not be ignored. A sum of £1000 was granted immediately and, on the main issue of the continuing support of the observatory, the Home Secretary announced that a commission to look into the whole matter was to be set up, a move which Piazzi Smyth greatly welcomed. Though in the event it was not to bring about any sweeping changes in Piazzi Smyth's time, it did certainly prevent the observatory from dying of neglect.

As if all these tribulations were not enough Piazzi Smyth had to fight a battle on yet another front. In 1871 a Commission of Enquiry into Scientific Instruction and the Advancement of Science had been set up to which the University of Edinburgh was asked to send a submission. In his capacity as Astronomer Royal for Scotland Piazzi Smyth was also expected to submit a report on the scientific work of the Royal Observatory. The university's submission involved a survey of the teaching courses provided by the various professors and of the student numbers in their classes. The preparation of this report triggered off a major dispute between the university and Piazzi Smyth. As soon as the university began to look into student attendances, it became obvious that very few students took the astronomy course—music was the other subject in the same category—and the university, anxious to satisfy the commissioners, looked for ways and means to cure this deficiency.

After the reforming Universities (Scotland) Act of Parliament of 1858 the University of Edinburgh had begun publishing an annual *Calendar* where the syllabus in practical astronomy, a straightforward summary of the positional astronomy course, printed alongside other courses, aroused no comment. After the first few years of teaching, in Piazzi Smyth's early days in office, it was evident that there was little demand for a full astronomy course among undergraduates. Nevertheless Piazzi Smyth kept the course open and delivered his lectures whenever any interested student came forward. He found that he could best be of service to serious candidates if he offered to teach them the particular subjects they wanted to know, and in 1871 he altered his *Calendar* entry to read:

> On account of the increasing amount of duties within the Royal Observatory Edinburgh on the one hand and the small number of students on the other, the Professor has limited himself to receiving, on the day announced in the University Calendar for the opening of the course, all matriculated applicants for Practical Astronomy, ascertaining their calibre and objects, and then advising and assisting such young men afterwards in their studies at periods throughout the session, according to circumstances, and hitherto gratuitously.

It was, under the circumstances, a very sensible solution. The type of person who from time to time came for tuition to Piazzi Smyth was interested in gaining practical experience in the handling of astronomical instruments. No systematic lectures were given; the student came up to Calton Hill in the evening by arrangement and practised with various instruments under the supervision of Piazzi Smyth or his First Assistant. No course fee was charged for this informal arrangement.

The university committee which were busy making a case before the London Commission did not take a sympathetic view of the special circumstances of the Chair of Astronomy. The university now had a new Principal, Sir Alexander Grant, successor of Sir David Brewster who had died in 1868, an ambitious man very different from Brewster under whom Piazzi Smyth had operated perfectly happily. As Principal of the University Grant was a member of the Observatory Board of Visitors, and was thus well aware of Piazzi Smyth's other duties in his capacity as Director of the Observatory. The findings of the university committee which were published in the local newspapers included details of professional salaries and gave the impression that the salaried Chair of Astronomy was held by a professor who had no students and by implication did no work. Piazzi Smyth's own report to the Commission of Enquiry into Scientific Instruction and the Advancement of Science, in which he laid out the unique position of his dual post, was ignored in the Principal's report to the same body. He protested to the Principal but got no satisfaction.

The crux of the matter was that while Piazzi Smyth looked upon the Royal Observatory as a national institution, like the Royal Observatory at Greenwich, under the charge of the Astronomer Royal for Scotland whose university Chair was quasi-honorary, the university regarded the Astronomer Royalship as an appendage to the Chair and saw the observatory only as a possible source of material for student instruction. The fact that Piazzi Smyth's salary was paid through the university was to him an accident of history; to the university, however, it proved that the Chair took precedence over the government appointment. It must be admitted that in his interpretation of his academic position Piazzi Smyth had changed his mind over the years; as a young man he had preferred to regard himself on all occasions as Professor of Practical Astronomy in the University of Edinburgh.

Piazzi Smyth did not lie down under what he rightly considered the unjust impression given by the university's report. He protested in the strongest terms to the Board of Visitors and made sure that his correspondence with the University Court was printed and circulated under the heading 'Recent Attempts to alienate the Astronomer's salary and destroy his character before the Public as an honest and working man.' In his annual report for 1873 he argued that the simplest and best solution would be to separate the two posts, free him from the 'revolutionary and in spirit red-republican claims of the University Court' and allow him as a government servant 'the leisure and peace of mind wherewith to accomplish his proper astronomical work within the Royal Observatory'. However, at the request of the university representatives on the Board of Visitors, who included Sir Alexander Grant, Piazzi Smyth was induced to withdraw his demand to have the university Chair separated from the government post; but he made no change in his teaching arrangements which even Sir Alexander no longer dared to oppose.

A last episode in this series of troubles, and one which Piazzi Smyth undoubtedly regretted in later life, was his resignation from the Royal Society of London in 1874. The circumstances could not have been more trivial, but under pressure from some of his fanatical pyramid disciples he found himself caught in a trap from which he could not escape. The innocent and unwitting agent in the affair was none other than James Clerk Maxwell who, in a famous address to the British Association for the Advancement of Science at their Bradford meeting in 1873, made a reference to the Egyptian cubit. This lecture, on 'Molecules' is one of Clerk Maxwell's most brilliant and most celebrated. It expounded two topics, the dynamical theory of gases, and the extraordinary fact of the identical nature of matter in all parts of the universe. On the second of these he described how the spectroscope had shown not only that all atoms of the same element here on Earth emit light of the same wavelength but that those in the Sun and the stars do so as well, even though the stars are so distant from each other that no material thing can ever have passed from one to the other. A consequence of this was that the imperishable atom could be used as a standard of measure.

To illustrate the point he made an analogy—and here came the offending words:

> From a comparison of the dimensions of the buildings of the Egyptians with those of the Greeks, it appears that they have a common measure. Hence, even if no ancient author had recorded the fact that the two nations employed the same unit as a standard of length, we might prove it from the buildings themselves.

The passage ended:

> A molecule of hydrogen, for example, whether in Sirius or in Arcturus, executes its vibrations in exactly the same time. Each molecule throughout the universe bears impressed on it the stamp of a metric system as distinctly as does the metre of the Archives of Paris, or the double royal cubit of the Temple of Karnac.

Needless to say the question of whether the Greek, Egyptian or any other cubits were the same was entirely immaterial to the argument, as Piazzi Smyth well knew. In fact when A S Herschel and George Maclear, Sir Thomas Maclear's son, visited the Smyths in 1872 they presented him with a small diffraction grating, the first one which Piazzi had ever seen, and Herschel told him about the possibility of using wavelengths of light as standards of length which could be related directly to the spacings of the grating. If further explanation were needed, Piazzi Smyth had Clerk Maxwell's own words, in a letter which was probably in response to one from Piazzi Smyth asking him where he had got his information regarding the cubits. Clerk Maxwell's letter dated 29 September 1873, repeats his discussion with such clarity that it is well worth giving it here in full:

> The statement in my lecture on molecules relative to the Greek and Egyptian measures of length was derived from the abstract of Lieut. Col. A. R. Clarke's paper on comparison of standards of length etc. with a preface and notes on the Greek and Egyptian measures of length by Sir Henry James, Proc. R. Soc. June 19, 1873. The first conclusion there stated is as follows. As regards the common cubit we have the statement of Herodotus that the Egyptian cubit was equal to the Greek cubit, that of Samos, and we learn from the measurements of the

Hecatompedron of Athens by Mr Penrose, that the Greek foot was equal to 1.013 foot or 12.156 inches and consequently the Greek cubit was equal to 1.520 foot or 18.240 inches. Then follow statements that certain dimensions of Egyptian buildings are round multiples of this cubit.

If my copy of your Great Pyramid researches had been here I should have been able to ascertain whether these statements were or were not consistent with yours. The cubit in which you take most interest is, I know, a longer one, but you admit the existence of a shorter one for common use and this may, or may not, be the same. So much for my authority which I was careful not to expose by giving numbers, as any errors in these numbers would have destroyed the illustrative value of the statement. Of course I might have compared the yards used in Britain and in the United States. The undoubted fact is that men have endeavoured at various times to construct rods which shall be as exactly as possible of the same length as a certain rod, and that these rods, or buildings measured with these rods exist in various places, and that it is an evidence of a desire of accuracy and just dealing on the part of the constructors.

Sir John Herschel's comparison to a manufactured article is not quite exact because uniformity in manufactured goods is often an accident depending on the mode of manufacture by machinery and is not an essential part of the design except in such things as Whitworth's bolts, screws etc. Whereas in weights and measures and coins, uniformity is of the essence of the use of the thing.

So far I think you will agree with me; at least your writings excite no opposition in me on all these points. Nay, further I am not prepared to differ from you on the practical value of Earth commensurability in standards or on the attainments of this either by the builders of the Great Pyramid or the constructors of the French metre.

My statement is that *if* we ever can measure the weight of a molecule of hydrogen, we shall *then* have a standard more permanent than any planet or heavenly body, but not till then, that is, not till what is now a merely conjectural estimate is converted into a well established physical constant. This molecular standard may possibly then be employed by those who expect that the authority of their numerical statements will outlast the planet on which they live.

Yours very truly
J. Clerk Maxwell

Having received this fully explanatory letter there was no longer any reason for Piazzi Smyth to feel aggrieved that his cubits had been overlooked and no question that he had not understood the important point being made by Clerk Maxwell. It is impossible to imagine that he would have made an issue of the reference to the cubit had he not been taxed by some of his ardent but ignorant pyramid supporters for allowing an 'erroneous' statement to appear in print 'with the authority of the Royal Society'. It was put to him that he, as a Fellow of the Society, had a duty to protest.

His protest took the form of a short paper criticising Sir Henry James with whom he had already quarrelled on the question of pyramid measures and whose paper mentioned by Clerk Maxwell was held responsible for the 'error'. Piazzi Smyth sent his paper, entitled 'On the length of the side of the base of the Great Pyramid', to the Royal Society on 27 October 1873, with a covering note saying that it was 'necessarily controversial'. He began by referring to James' paper 'having led Professor Clerk Maxwell into a serious error in an Egyptian allusion ventured by him in his otherwise most admirable address on "Molecules" before the British Association'. He recapitulated his own researches into the various cubits and then proceeded to attack James for an 'alternate selection of some and concealment of other data to suit successive hypothetical views'—these were, incidentally, the very same accusations which Sir James Simpson had made against Piazzi Smyth in the great Royal Society of Edinburgh row!—which 'has led both the Royal Society and Professor Clerk Maxwell unwittingly to confirm the act'.

A brusque letter from G G Stokes of Cambridge, then Secretary of the Society, informed Piazzi Smyth that his paper had been rejected by the Publications Subcommittee on the grounds that it was 'not suitable for a public reading before the Society'. Piazzi Smyth considered a reponse for two weeks before writing on 7 February to the President of the Royal Society, Sir Joseph Hooker:

> With reference to my letter of last October to the Secretary of the Royal Society forwarding a short paper entitled "On the length of the side of the base of the Great Pyramid" and with reference also to the rejection and return of the said paper on the plea by a subcommittee of its not being of a nature suited for a public reading before the Society, I beg to say that having thus failed in all that I can do to open the eyes of the Society as to whether they are seeking "accurate measuring, truth stating and justice doing", or the exact opposite thereof, in researches concerning the most ancient and exalted monument of intellectual and religious man on the face of the Earth, there is nothing now left me but to come out of the Royal Society, as I do hereby, resigning my Fellowship therein and hoping that you, Sir, at least, will not consider my reason for withdrawal "not suitable for a public reading before the Society."

The words 'accurate measuring, truth stating and justice doing' came from the last lines of Clerk Maxwell's discourse. The letter was intended only as a threat, and Piazzi Smyth anticipated that when it was read before a meeting of Fellows, they would repudiate the

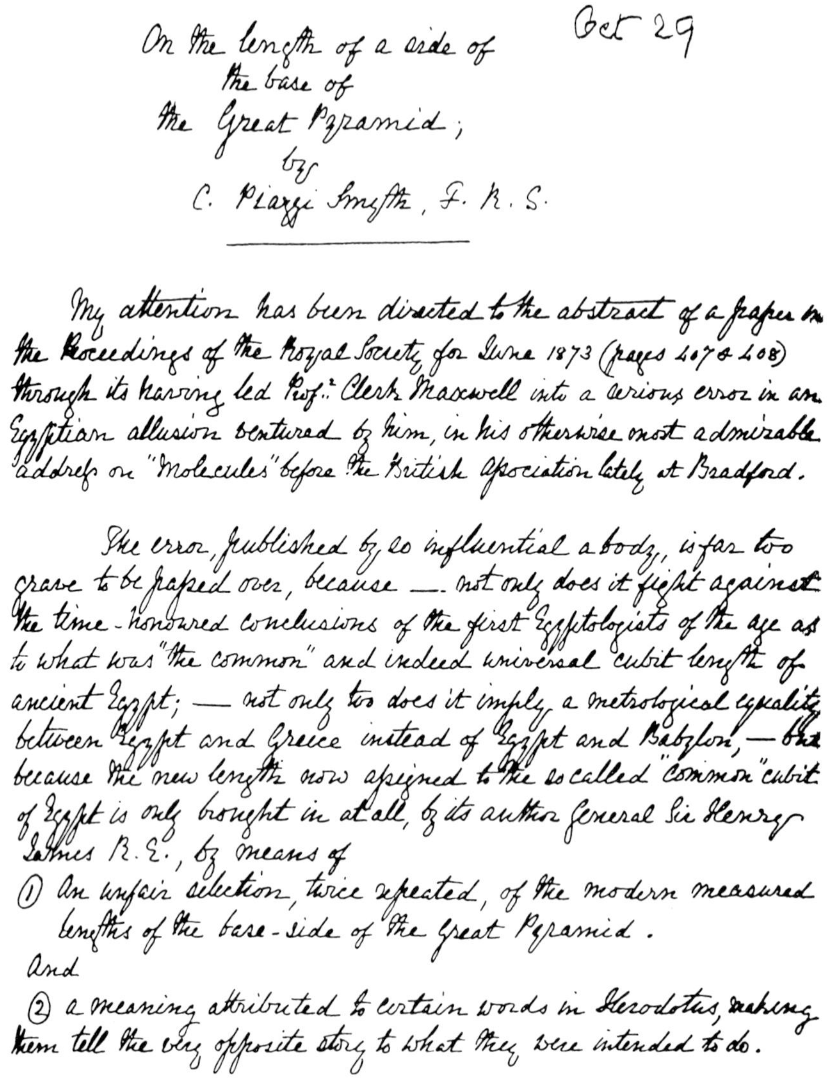

Oct 29

On the length of a side of
the base of
the Great Pyramid;
by
C. Piazzi Smyth, F. R. S.

My attention has been directed to the abstract of a paper in the Proceedings of the Royal Society for June 1873 (pages 407 & 408) through its having led Prof.r Clerk Maxwell into a serious error in an Egyptian allusion ventured by him, in his otherwise most admirable address on "Molecules" before the British Association lately at Bradford.

The error, published by so influential a body, is far too grave to be passed over, because — not only does it fight against the time-honoured conclusions of the first Egyptologists of the age as to what was "the common" and indeed universal cubit length of ancient Egypt; — not only too does it imply a metrological equality between Egypt and Greece instead of Egypt and Babylon, — but because the new length now assigned to the so called "common" cubit of Egypt is only brought in at all, by its author General Sir Henry James R.E., by means of

(1) An unfair selection, twice repeated, of the modern measured lengths of the base-side of the Great Pyramid.

And

(2) a meaning attributed to certain words in Herodotus, making them tell the very opposite story to what they were intended to do.

First page of Piazzi Smyth's controversial paper, in Jessica's hand, sent to the Royal Society (ROE).

Subcommittee and refuse to accept his resignation. Instead, his letter was laid before the Council and the resignation was accepted. The outcome of his rash offer to resign left him shocked and furious. This was an entirely different result from that which had followed the same threat and tactics at the Royal Society of Edinburgh when the Council had convened a special meeting and the President had personally hastened to his house to coax him to withdraw his resignation.

Unfortunately, Piazzi Smyth was not satisfied to leave matters alone at this stage. He prepared and printed a pamphlet containing the full correspondence with the Royal Society which was sent out to all Fellows and to many others outside. The affair was, not surprisingly, reported in the Press, mostly in a critical or mocking tone. His eccentric views which had to a large extent been forgotten except in the world of his devotees were now exposed in all their folly to the Fellows of the Royal Society and to the public at large.

Among Piazzi Smyth's scientific friends and colleagues there was almost unanimous agreement that he had seriously erred in imputing base motives to Sir Henry James, and that he had behaved foolishly in offering his resignation. At the same time they sympathised with him and felt that the Royal Society might have been a little less quick in taking him at his word. Piazzi Smyth's old master, Maclear, who could speak frankly like a father, wrote to him immediately:

> I regret, most deeply regret, your withdrawal from the Royal Society, and the tone of the language which, seemingly, encouraged the Society to dispense with your Fellowship. Such expressions as "knowingly neglected", "misrepresented" etc. imply charges of dishonesty for which there was no need. Sir Henry James is trying to get at a certain ancient measure of length. The enquiry involves experiment. He does not venture to deal with the "Divine Inspiration" theory—you do—and you believe the structure was perfect.

The letter ended: 'A half a cubit is not large enough to justify a wrangle, therefore pray, my dear fellow, forgive my touching on a matter wherein you must admit you are an enthusiast.'

Most of his friends wrote in the same vein. Pritchard, Professor of Astronomy at Oxford, a man of great distinction and like so many others a good friend of Piazzi Smyth, said: 'I have read your pamphlet with great concern. If this be the whole that occurred, reasonable men must say that you have been unhandsomely treated. I think that your resignation ought not to have been so easily accepted; you ought to be reinstated.' Robert Mallet, the geologist who had taken a strong scientific interest in Piazzi Smyth's pyramid researches, said: 'I think you have been treated cavalierly, but your resignation was as it seems to me an unfortunate step.' J R Napier, the engineer, summed the whole matter up when he wrote to Piazzi Smyth: 'I am sorry that you thought it necessary to retire from the RS for the reasons given in your pamphlet. I take leave to offer a criticism. It does not appear to me that the Royal Society *could* publish the paper in the form you sent it. If you had sent them a paper giving the measures with your explanation of the Samos cubit without even mentioning Sir Henry James or certainly without imputing motives to him, I think the Society could not have refused to publish your remarks. The Council have evidently checkmated you by accepting your resignation according to their rules without reading your reasons. I cannot help thinking that you have wronged yourself by resigning and in the publication of the correspondence. I am very sorry, however, that it should have happened.'

Some of his friends defended him publicly. Among them was the faithful H L Smith of Hobart College in the United States who wrote a long letter to the New York *Herald Tribune* in which almost the entire contents of the pamphlet were reproduced with his own exaggerated denunciations of the Royal Society. This was greatly appreciated by Piazzi Smyth who printed the letter in the next edition of his book *Our Inheritance in the Great Pyramid*. The Petries also supported him. The son, Flinders, now a young man of 21 and already taking his first steps into archaeology surveying earthworks and stone circles in the south of England, wrote to the *English Mechanic and World of Science*, a popular scientific journal, defending Piazzi Smyth against the journal's account of the Royal Society affair.

Religious personal friends expressed their sympathy to Jessica: 'How cruel it must fall on you and Mr Piazzi to be so treated.' 'It is well that the Professor is a man of great patience. What a good thing there is no Inquisition in England or *somebody* might meet with poor Galileo's fate!' To the fanatical pyramidology disciples the affair was a triumph; they gleefully boasted at what they thought was the defeat of the arrogant and godless scientists.

Piazzi Smyth was never re-elected to the Royal Society, though 20 years later, in his retirement, he received regular grants from it towards his researches into solar spectroscopy. The whole episode was as sad as it was unnecessary. Not only did it draw ridicule on his head; it also put a barrier between him and the influential scientists of the 'establishment'. By no longer publishing his scientific papers with the Royal Society of London much of his excellent researches in spectroscopy received less notice than they deserved.

The episode with the Royal Society was the last time that Piazzi Smyth discussed the pyramid in scientific circles. However, as his private correspondence shows, his personal following in what had become a cult continued to grow. It was as if he lived two independent lives, one as an orthodox scientist, the other as a religious devotee of the pyramid.

Chapter 11

A Wet France and the Rainband

Following his successes with spectroscopes of moderate size, Piazzi Smyth planned to use instruments of high dispersion particularly for solar work. He wanted to resume the problem of discriminating between lines of solar and terrestrial origin in the spectrum of the Sun, while at the same time he was becoming more and more attracted by the field of spectrum analysis of laboratory sources. As Alexander Herschel said of him, the possibility that the chemical composition of the Sun could be read in its light's prismatic spectrum acted on him 'like a spell'. Ångström's map of the normal solar spectrum had identified 14 elements as definitely present in the Sun but these accounted for only a few hundred out of the thousand lines which he had listed. There were many more chemical elements to be discovered if a match could be established between solar lines and those in the spectra of substances made luminous in the laboratory. Piazzi Smyth was fascinated by the thought that with a set of good vacuum tubes each containing a residuum of some particular pure gas a spectroscopist in his laboratory could 'enquire at any moment he pleases into the physical constitution of what makes up near half the Universe'. For Piazzi Smyth the mysterious spectrum of the Aurora also beckoned. He resolved to construct large spectroscopes and to teach himself how to produce artificial light sources in the laboratory.

Before embarking on his new spectroscopic investigations Piazzi Smyth studied the catalogues of the leading spectroscope makers and of the electrical suppliers of the day. He filled pages with sketches and calculations using his own straightforward method of designing the prisms for his instruments: the makers' specifications of dispersion and deviation angle for prisms of various angles and types of glass were scaled and combined on paper to produce the desired dispersions and deviations without any theoretical computations. The fact that he had no profound knowledge of theoretical optics was no bar to his capability of putting together some very efficient spectroscopes, large ones with fluid prisms as well as small direct-vision instruments. His favourite optician was J Salleron of Paris whom he decided to consult first.

Armed with his specifications Piazzi accompanied by Jessica travelled to Paris in June 1875. He found the city much changed since their last visit in 1869 with 'lamentable destruction of the Tuileries and other monuments by the Commune' but otherwise, to his approval, 'orderly and subdued by moral control'. They stayed at the Grand Hotel in the Boulevard des Capuchins where they had 'perfect quiet in the breakfast room, tranquil seclusion in our own, and good wine generally'.

Salleron, 'sadly old looking', had an impressive establishment with several skilled assistants. Piazzi Smyth discussed his plans with him returning several times while Salleron worked out the details of the proposed spectroscope to his entire satisfaction. The spectroscope which was to take a year to complete was ordered. Paris was at that time full of opticians and advanced instrument-makers of all kinds. The Smyths naturally visited Secretan's, the opticians responsible for the 31-centimetre Paris refractor and saw the workshops of makers of chemical apparatus, induction coils, chronographs, vacuum tubes and other modern laboratory necessities. Particularly recommended by Huggins was H D Ruhmkorff, the well-known maker of induction coils, 'a grand grey-haired old man'. Piazzi Smyth unfortunately could not afford for himself any of Ruhmkorff's coils, priced at £460, but he had a commission to inspect them and also vacuum tubes and other items in a shopping list of a more affluent friend, Mrs Mitchell, an amateur physicist and astronomer and the wife of a Scottish Member of Parliament.

The Smyths, of course, wished to visit the Paris Observatory to pay their respects to the great Le Verrier who, having been removed from his post in 1870, had just been reinstated as Director by Thiers. Piazzi had met Le Verrier before and had frequently corresponded with him, but all the same it was with some trepidation that they presented themselves because of Le Verrier's reputedly difficult personality. At the observatory—'the only building in Paris not inscribed with "Liberté, Egalité, Fraternité", but simply "Observatoire"', as Piazzi Smyth approvingly noted—he and Jessica found M and Mme Le Verrier and another visitor, H C Russell, the Director of Sydney Observatory, known for his visual and later photographic work in the southern Milky Way. Le Verrier was willing enough to show them his new 48-inch silver-on-glass reflector but did not exert himself any further, leaving his visitors to find their own way home in a great storm of lightning and rain.[1]

A warm reception awaited the Smyths when they went to the offices of the journal *Les Mondes* and its editor, the Abbé Moigno. Moigno was a staunch champion of the pyramid theories about which he had published several articles in his journal. His translation of Piazzi Smyth's *Our Inheritance in the Great Pyramid* into French had just appeared under the title *La Grande Pyramide, Pharaonique de Nom, Humanitaire de Fait, ses Merveilles, ses Mystères et ses Enseignments*. Outside his editorial duties Moigno, now a Canon, led a simple existence in a village near St Denis not far from Paris, where he was looked after by his niece and grand niece. Here the Smyths after Sunday lunch attended Vespers in the cathedral and the Abbé's service in the little Catholic chapel of the Hospital for Incurables where they listened to Moigno's sermon on St Peter. It was a strange but very genuine friendship which bound the evangelical fundamentalist with the humble Catholic priest.

In pursuit of his curious interest in burial places Piazzi Smyth visited the cemetery of Père La Chaise where he admired and recorded the chapel-like tombs and watched the crowds with their watering pots tending the plants and flowers on the graves. The tombs of the great were not forgotten: he visited the crypts of kings back to Clovis, Pepin and Charlemagne in the Cathedral of St Denis, the patron saint of the French people.

A visit to Paris would have been incomplete without seeing the Louvre and in particular the Egyptian gallery which Piazzi surveyed with a critical eye, finding many faults in the catalogue as regards dates and descriptions. At the Conservatoire des Arts et Métiers, the new Museum of Science and Arts which Piazzi Smyth had not previously seen, there were 'miles and miles of glass cases' full of models and instruments. To him it was a grave, not a living exhibition, which led him to comment that it is 'better to teach oneself the use of one chisel than see innumerable glass cases full of carpentry', a sentiment which aptly describes his own approach to experimental science.

The downpour through which the Smyths had been unceremoniously allowed to stumble back to their hotel after visiting the observatory ushered in an unusually rainy spell, contrary to the prediction of fine weather previously issued by the Paris Meteorological Department on the basis of high barometric readings over England. The weather got wetter by the hour and was no better in London when the Smyths arrived there some days later, though the barometer remained high. How could this be, Piazzi Smyth asked himself?. His pocket spectroscope, which he was in the habit of carrying around with him showed, in whatever direction of the sky he pointed it, a broad dark band on the red side of the well-known yellow sodium absorption lines; in his little low-dispersion instrument the band merged into the sodium lines which thus appeared to be abnormally intensified. The band was so strong as to be the dominant feature of the whole spectrum and though undoubtedly telluric in origin it seemed very different from the normal atmospheric effects which increase conspicuously when seen in the spectrum of the Sun at sunset. Piazzi Smyth first thought that the phenomenon represented some obscuration by the London smoke, but he continued to see it in the smokeless countryside from the train all the way north to Scotland. The band could still be seen next day in a clear blue sky, which was followed the same evening by rain which continued for several days.

Piazzi Smyth reported his observation to *Nature* and continued watching the spectrum of the sky, as did his friend and colleague P G Tait. Piazzi Smyth was soon able to deduce that the band was identifiable with 'warm rain in an easterly wind and under a high barometer'. He saw the potential of what he called 'rainband' observations for weather forecasting—not as a replacement for the barometer and the hygrometer, but as a supplement to these. The barometer indicated atmospheric pressure, the hygrometer or wet and dry bulb thermometer gave the percentage of moisture in the immediate vicinity of the instrument; but the spectroscope, analysing the moisture through many layers of atmosphere, gave a more direct picture of the average degree of moisture and could presage rain before it reached the nearby stratum of atmosphere. Piazzi Smyth, or rather Jessica who already kept a meteorological journal at their Edinburgh home and also had her own portable set of meteorological instruments, thenceforth added daily observations of 'rainband intensities' on a scale of 1 to 10.

Piazzi Smyth's were not the first observations of atmospheric water vapour lines in the vicinity of the yellow D lines. A group of lines between the two D lines had been discovered in the solar spectrum much earlier by the American spectroscopist J P Cooke who recorded a marked increase in their strength when observed under the very humid warm conditions of the season known in New England as 'Indian summer'. His results, in which he correlated the effect with hygrometer readings, were published in the *Proceedings* of the American Academy in 1866. However, the two independent researches differed in that Cooke had been concerned with identifying individual atmospheric water vapour lines in the spectrum of the Sun while Piazzi Smyth was attracted to the potential of the whole water vapour band seen under low dispersion as a weather indicator.

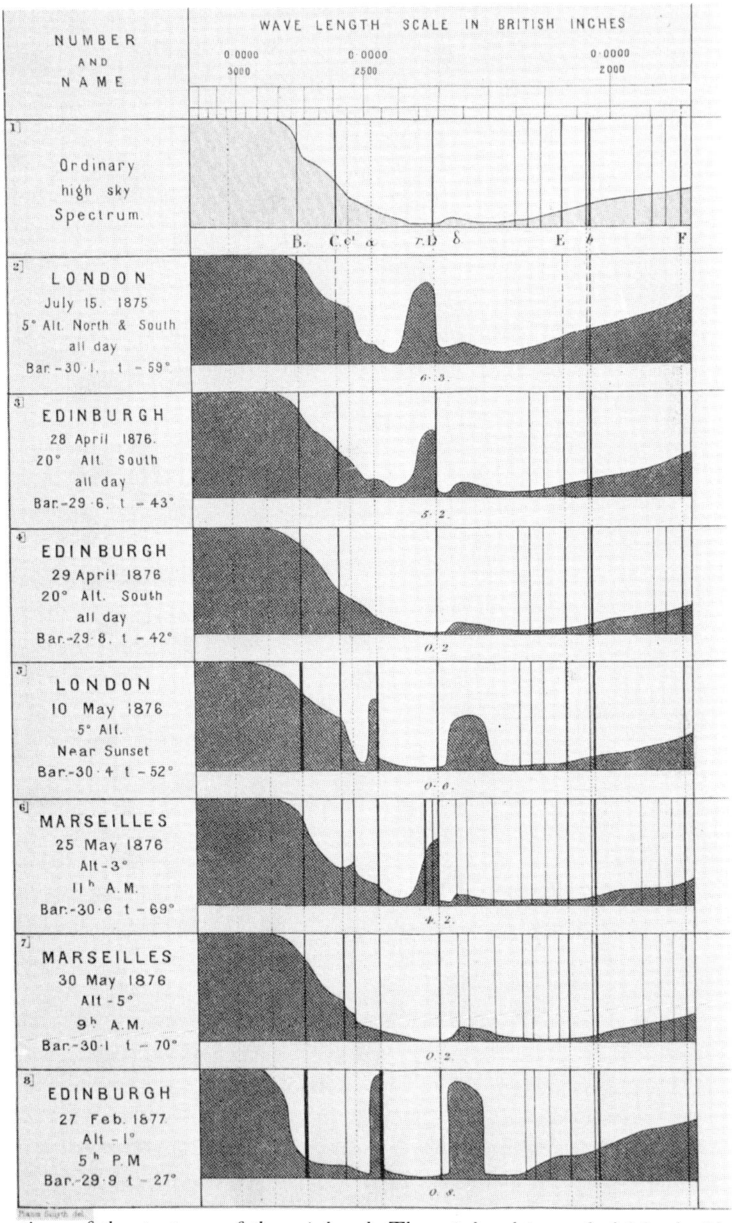

Observations of the spectrum of the rainband. The rainband is marked 'r'; the 'dry air band' δ monitored the effect of the altitude of the Sun on the sky spectrum (CPS, MNRAS vol. 32).

After two years' experience Piazzi Smyth could confirm the usefulness of his observations of the 'rainband' by correlating them with the detailed meteorological data that came his way from all over Scotland. He contended that while the barometer is effective in foretelling rain in winter, because rain is then brought on by violent south-west storms, it is less so in the case of spring and summer rains which come from the east or south-east. Insomuch as changes of the weather begin high up in the atmosphere, something he had experienced from the Peak of Tenerife, it was not surprising that the spectroscope could discover moisture at high levels sooner than the hygrometer which sampled only a small volume of air around the instrument. Piazzi Smyth published his findings in the *Edinburgh Astronomical Observations* for 1878, together with instructions on the method to be used for the recording of the rainband and tables for the reduction of the observations to a standard norm for temperature and thickness of air traversed.

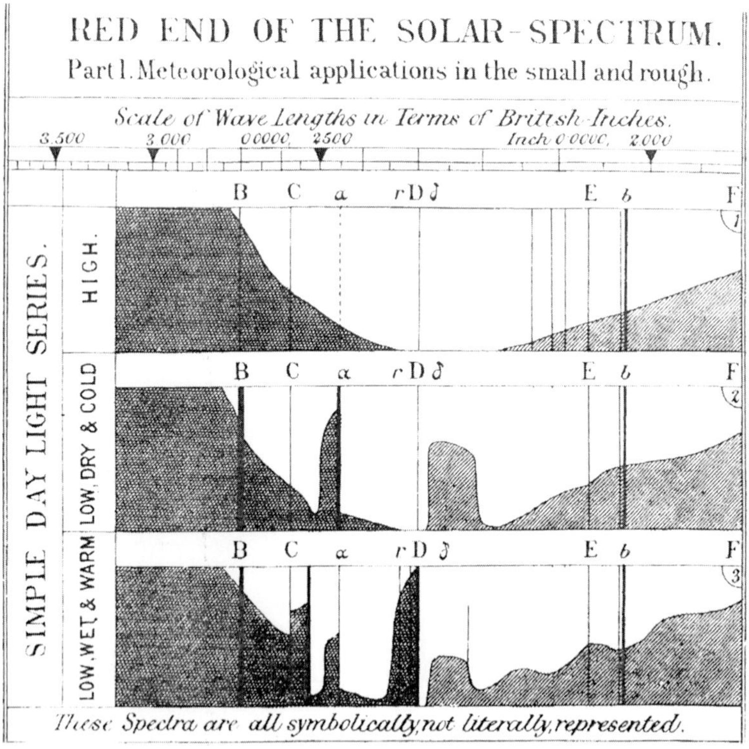

Piazzi Smyth's guide to the rainband (CPS, Astronomical Register vol. 6).

The rainband had many devotees among them C Michie Smith, a former student of Tait's who became an astronomer at Madras and later set up the Kodaikanal Observatory. He found that in India a rainband seen by day was always followed by thundery rain in the evening. The rainband was also popular with amateur meteorologists one of whom, Rand Capron, an experienced auroral observer and spectroscopist, made a plea for its systematic use for weather forecasting. Hilger were prompted to produce a number of 'rainband' or 'meteorological' spectroscopes to Piazzi Smyth's design; these were low-dispersion direct-vision instruments, four inches long and three-quarters of an inch in diameter and easily portable. Hilger's hopes that they would be widely sold as a sort of universal weather forecasting device were not fulfilled because they were not as easy to use as Piazzi Smyth with his experienced eye had found them. An observer had first to become familiar with the normal spectrum of the sky in the region of the sodium D lines before he could judge whether there was additional darkening of the spectrum to the red side of this feature. The occurrence of the rainband was a phenomenon which had to be interpreted in conjunction with the barometer and hygrometer; and a rainband spectroscope was a scientific instrument for the professional rather than a handy gadget which would give quick answers to the ordinary user. The rainband spectroscope did not catch on with the general public, though the London optician and instrument-maker Louis Casella tried to revive interest in it some years later. Among the spectroscopists who took up seriously the study of the rainband were Janssen, Cornu and Thollon in France, and L Bell in the United States who in 1882 made an improved version of the rainband spectroscope which he claimed had a high degree of success in weather forecasting.

Piazzi Smyth's rainband work signifies an early attempt to add a quantitative element to the indefinite science of meteorology and as such deserves recognition. With the same purpose he also carried out experiments in cloud photography, a field into which he was the very first to venture. He believed that photographs of different cloud formations were greatly superior to verbal descriptions, and that the study of systematic records of these could not fail to provide a scientific basis for weather forecasting. He designed a special camera to produce negative photographs of clouds on small plates. He described his work in an address to the Edinburgh Photographic Society in 1876 which was published in the Society's *Proceedings*. The cloud camera earned for him the Society's Silver Medal in 1877. Piazzi Smyth returned to his cloud photography in his old age by which time, almost 20 years later, the scientific fraternity had taken up the same technique of cloud observation.

The Smyths went back to Paris a year later to inspect Salleron's 'great spectroscope' and were delighted to find it well up to expectation. It was an instrument of substantial dimensions, consisting of a pair of 30-degree flint prisms and two lenses with apertures of 2.25 inches and focal lengths of 15.5 feet. Piazzi Smyth found it far superior to the one supplied by Grubb for the Edinburgh equatorial, showing 'lines innumerable' in the spectrum of the sky. Five alternative prisms were also supplied, as well as a longer-focus objective and a heliostat for the purpose of directing sunlight into the spectroscope. Salleron made a few final adjustments while the Smyths made a tour of southern France and visited the observatories of Toulouse and Marseilles. On their return to Paris they found that the spectroscope was ready for shipping to Edinburgh and the happy transaction was celebrated 'with much clinking of glasses'.

The tour of southern France is another of those recorded in lively detail of word and picture in Piazzi's diaries. Piazzi was in the mood to enjoy the journey which began in the evening at the Orleans railway station in Paris:

> Off and away into the evening along the banks of the Seine on the smoothest of railways. Away and away over meadows and gardens amid fortification-disturbed land. Away and away through the growing shades of night with Venus blazing in the west when daylight disappeared. Away and away through the night, and when Venus has set Jupiter slopes towards her place with new stars; all night long the stars are so bright. About 2 a.m. the Moon is seen golden; afterwards when higher, silver. Then comes dawn and after passing Limoges in the cool air of the morning the valley mist makes a gigantic river, setting off lines of poplars in the valley, and warm coloured hills.
>
> From Limoges to Ville Franche and almost to Toulouse the country is La Belle France; well cultivated below, rising above into hills and table lands coronetted with rocky precipices and then crowned with old castles, churches, chateaux. Some magnificent sweeps of river and cliffs reflected in the water. The highest part of the country is 1200 feet by aneroid.

The last line refers to the portable barometer–altimeter, part of Piazzi Smyth's arsenal of meteorological apparatus which he carried with him everywhere and which now included his own special device, the rainband spectroscope. Somehow he managed to write his notes, sketch the scenery and keep his hourly meteorological record.

The observatory at Toulouse had as Director the renowned mathematician F F Tisserand who in 1883 was to succeed Puiseux at the Sorbonne where he wrote his great four-volume treatise on celestial mechanics. In Toulouse he had two assistants, the senior of whom, H J Perrotin, was well known for his discoveries of minor planets. Though the observatory possessed several meridian instruments by well-known makers, its work was mainly concerned with extra-meridian astronomy for which there was a new large 80-centimetre reflector in the grounds. Tisserand, 'a little smiling sailor-like man, obliging and merry' made the Smyths welcome and showed them the planets and the Ring Nebula through the reflector. It was a Sunday evening, and Tisserand had spent the day preparing the next day's lecture to be delivered at the university at 8.30 in the morning. Piazzi Smyth was most impressed. He had drawn up in advance a list of questions for Tisserand, and, being now somewhat deaf, he had written these down leaving spaces for Tisserand's replies. After some questions on the history of the observatory, staff and budget, he asked: 'Are the meteorological observations an obstacle to the work of the Observatory?' and 'Does the educational work interfere with the work of the Observatory?' If Piazzi Smyth was expecting ammunition in his complaints about his own meteorological duties and in his quarrel with the University of Edinburgh, he was disappointed, because the answer was a plain 'No' to both questions! Indeed Piazzi Smyth could hardly help noticing that Tisserand with the same staff as he himself had in Edinburgh was perfectly happy with his lot.

At the Marseilles Observatory there was much of interest to see. It was new, in fact not yet entirely completed, with as yet no entrance road or dwelling house for the Director. Instead of the usual single, large building, there appeared to be a number of them spread around the open ground. Piazzi Smyth was interested to learn that the observatory had been placed in Marseilles partly because of its usefulness for longitude determinations of Algeria. The staff consisted of the Director, E Stephan, and three assistants. The 'great equatorial' was the 80-centimetre silver-on-glass reflector made by Foucault and Eichens which had been originally installed in Paris but which was moved to Marseilles in 1864. Stephan explained to Piazzi Smyth how the mirror was silvered once a year by himself. Stephan, 'amiable and obliging', showed the Smyths the reflector and an excellent 9.5-inch Merz refractor. The quality of these instruments was most impressive and indicated the more professional attitude of the French astronomers to their instrumentation than that of their British counterparts who were still dominated by an amateur approach. It was this latter attitude that cost Piazzi Smyth dear and lost the Edinburgh Observatory its one chance in a century of leading British astronomy with a telescope of the first rank. On one matter only was Piazzi Smyth ahead of his French colleagues; that was in his conviction that it was a waste of time erecting observatories so close to smoky cities. He was reluctant to accept Stephan's claim that Janssen had found the astronomical definition no better in the Himalayas than he did in the Indian Plains. On Tenerife, he said, the result had been quite different.

One morning in Marseilles Piazzi's spectroscope foretold rain and, as he expected, the rain came down and continued steadily all afternoon and evening. In the hotel the dining tables laid out in the open air had to be rapidly dismantled, and Piazzi wryly added a footnote to his meteorological observations: 'Hotel keeper had chosen today for redding the regular table d'hote saloon, expecting fine weather and to dine his clientele in the open—and now, confusion! He not having been skilled in the rainband interpretation!'

All the way back to Paris by train, a total travel time of 20 hours, Piazzi Smyth continued with his rainband observations during daylight, as often as two or three times per hour. Each observation, which also included temperature and barometer readings, was accompanied by a tiny coloured picture and a written description of the scene. This feat illustrates yet again Piazzi Smyth's never-flagging tenacity, not to mention the speed at which he could write and draw, a talent which Mrs Maclear had marvelled at when he was a 16-year-old youth in South Africa.

Before finally leaving France Piazzi Smyth had what to him was a most important call to make, namely to the astronomer Janssen who had succeeded in observing the spectrum of water vapour by looking at the light of distant gaslights through a long tube of steam. This was an experiment which had been tried for the first time in Edinburgh by J D Forbes with only limited success 30 years earlier. Piazzi Smyth felt that Forbes' pioneering experiment, 'a most important physical discovery in its day and by the most eminent physicist as well as exact experimenter Scotland has ever produced', had never been properly recognised. In 1871 Piazzi Smyth, assisted by Jessica and his two assistants from the observatory, had carried out a similar experiment in Edinburgh which consisted in looking at the spectrum of a gaslight through jets of high-pressure steam from railway engines in St Margaret's railway yards, obligingly arranged by the railway manager. The effect had been to shorten the spectrum at the red and blue ends, but neither Piazzi Smyth nor his observers were able to recognise any of the special features reported by Janssen.

Janssen was at this time still working from his private house near Montmartre where the Smyths were received kindly by the Janssens and their daughter whom they already knew from the British Association Meeting in Edinburgh in 1871. However, Piazzi had to admit to being disappointed at not being able to learn any details of the steam spectrum. He had hoped to see a chart showing the positions of individual absorption lines, and to find confirmation of his 'rainband', but Janssen could only show him the same material which he had already published and his 'high and low sun' spectra. The latter showed the 'rainband' lines but 'the spectrum of water vapour is as unsatisfactory as ever; there is no existing drawing of it pure and simple'.

The Smyths enjoyed convivial company but rarely allowed themselves to deviate from their serious purposes. In Paris they made an exception. On the last evening they treated themselves to a final dinner in their favourite restaurant, Duval's near the Madeleine. As to many another returning visitor from France, London seemed a dreary place by comparison. On the cab drive to their hotel in Gloucester Road they were upset by the contrast between London's aristocratic squares and the rows of brick cottages where the population 'seems to have no hope of advancing themselves to any other than their pursuit of their low coarse and repulsive existence'. Yet a similar contrast could be found within yards of their own doorstep in Edinburgh; one suspects that the Smyths, like other worthy citizens of the Scottish capital, did not wander far from the New Town.

At the hotel in London, in spite of 'no end of blue velvet sofas in the drawingroom', the food and service were not up to French standards. 'The men waiters bring around the dishes with grimaces of smiling and subservience so awkwardly done that the duty is clearly unsuitable to Englishmen. There is nothing here like the system in France, the training and precision of the men waiters in the Grand Hotel and the propriety and pleasantness of the waitresses at Duval's. The sooner English waiters go and serve as sailors before the mast or go and colonise foreign lands and leave waiting to women, the better.' However, they were appropriately entertained in the homes of their numerous relatives and old friends including Henry Clerk, now retired but still busy in his retirement on improvements of gun recoils. Particularly interesting was a visit to William Huggins and his wife at their private observatory at Tulse Hill in the suburbs of the city. Huggins, one of the greatest astronomical spectroscopists of the age, had set up the observatory at his home in 1856 and had made pioneering observations in every field of stellar spectroscopy. He now had the assistance of his gifted wife, whom he had married only the previous year and who was to become recognised as his partner in their scientific labours for the rest of his life. Piazzi Smyth already knew Huggins who had visited him in Edinburgh on more than one occasion. The Smyths went by train out 'through the bowery roads and gardens of Clapham' and walked a long way from the station to reach Huggins' modest house, with its painted glass entrance door depicting Comet Halley on the Bayeux tapestry. Inside, the Smyths encountered the Huggins' dogs, a big yellow mastiff called Kepler and a little black terrier called Tycho Brahe. The dome, reached through the house, was carpeted and had plenty of space for laboratory instruments of all kinds—batteries, induction coils and a Leyden jar—as well as the Royal Society's 18-inch Grubb reflector. For his spectroscopes Huggins insisted on prisms made of very perfect glass, and Piazzi Smyth could see the result of this in the 'exquisite' definition of fine lines in the spectrum of sunlight thrown through a window from a heliostat. Piazzi Smyth, fresh from his visit to the modern French observatories and conscious of the splendid Parisian spectroscope which was already on its way to Edinburgh, wrote of Huggins in his diary, with just a touch of smugness: 'He has much liking for his old pieces of apparatus and for making up things himself and was once beyond everyone else in spectroscopy. But now even with Mrs H.'s assistance he must see to it that others do not pass him in refinements and fine mechanics'—a very mistaken impression! Huggins—who incidentally was six years younger than Piazzi Smyth—and Mrs Huggins were well able to hold their own against the rising generation of professional astronomers in the new astrophysical observatories and continued to make substantial contributions to spectroscopy for another 20 years.

An entire week was devoted to touring the exhibits of the extensive loan collection of scientific instruments in the South Kensington Museum, where some of Piazzi Smyth's pyramid instruments were on display including his measuring rods and a 'pyramid thermometer' graduated in Piazzi Smyth's own system, with a range of 0 to 250 degrees between the freezing and boiling points of water. Piazzi's brother Warington was on the committee of this large educational exhibition. There were also the London opticians: the well-known firm of Browning, who made a direct-vision spectroscope for Piazzi Smyth, and Adam Hilger, a firm less well known at that time, who had been recommended to Piazzi Smyth in the first instance by H C Russell of Sydney whom he had met at the Paris Observatory. Piazzi Smyth ordered a set of achromatic eyepieces from Hilger on this his first visit thus starting what became a most felicitous association with this firm. When some months later a Scottish friend, Sir Archibald Campbell, later Baron Blythswood, a wealthy amateur scientist, commissioned various scientific instruments from Hilger, the younger of the brothers, Otto Hilger, was dispatched to Campbell's residence, Blythwood House in Renfrewshire, to install them. Campbell writing to Piazzi Smyth was full of praise for Otto Hilger's superb talents with the result that on his way back to England Otto visited Piazzi Smyth in Edinburgh to discuss some new ideas regarding additions to his spectroscope. When Otto returned home to London he wrote appreciatively—with echoes of his native German—to his hosts: 'The principal object in writing to you is to thank you and your dear Lady for the kind reception you proved to me at my stay in the beautiful romantic Edinburgh which I shall never forget; the only way in which I can revenge myself is in getting your orders with which you favoured me truly and conscientiously executed.' The Hilgers certainly 'revenged' themselves with their excellent work. Over many years they produced several prisms for Piazzi Smyth's spectroscopes, first solid glass and quartz ones of various designs and later liquid ones.

Salleron's spectroscope and heliostat duly arrived in Leith right in the middle of the sitting of the dreaded Government Commission of Enquiry. Piazzi Smyth could distract himself with his wonderful array of new apparatus. His favourite among the new prisms became No 3, his 'Pyramid prism', made of white flint with an angle of 51° 51′, the magic slope of the π-pyramid. The time being now sunspot minimum with no displays of Northern Lights, the spectroscope was installed in the southfacing solar laboratory where the heliostat mounted outside the window was manually operated by Jessica while Piazzi looked after the spectroscope.

So many prisms were made by Hilger and Salleron that it is now impossible to keep track of Piazzi Smyth's various spectroscope arrangements. It would appear that the original spectroscope made by Salleron was never used in the exact form in which it left the maker. Piazzi Smyth liked experimenting with his prisms and using whatever arrangement best suited his particular purpose. Thus the so-called 'Aurora spectroscope' which Piazzi Smyth also used for his laboratory spectroscopic investigations which he mounted in his north-facing room on the top floor of his residence was described as 'constructed mainly by M. Salleron, Paris, to my design in 1876 but has since been much improved and strengthened', while the spectroscope used for solar work in Portugal in 1877 and 1878 was put together by himself with 'the several important parts furnished by, and altogether due to, the professional skill of M. Salleron, Paris, and Mr Adam Hilger, London'.

The Edinburgh Royal Observatory Commission eventually met in 1876, three years after William Thomson's intervention. Piazzi Smyth's first intimation was, as he recorded in his dairy (on 8 May 1876), a 'grand letter from Government announcing a Commission of Enquiry on Royal Observatory, Edinburgh, composed of Lord Lindsay, the Hydrographer, The Queen's and Lord Treasurer's

Remembrancer, the Professor of Natural Philosophy, Edinburgh, Mr Robert Anderson, Home Office, and self. Self to correspond and settle when. I write to them all accordingly. I will call on Mr Anderson (who will be Secretary) on Wednesday, perhaps too on Hydrographer'. Piazzi Smyth was setting out the very next day for France and it suited him admirably to call on these men on his way through London. To his dismay and to the embarrassment of the Home Office he discovered when he paid his visit to Whitehall that the original letter had been sent to the wrong Astronomer Royal! It was intended for Sir George Airy; as Piazzi Smyth wrote, 'I to be the witness, not the judge, for Edinburgh Observatory.'

The relations between Airy and Piazzi Smyth at this time had become tenuous. Sir George Airy had supported Piazzi Smyth cordially in the early days, most especially in the Tenerife venture. In J D Forbes' time the Airys were regular visitors to Scotland but as Airy wrote to Piazzi Smyth in 1866: 'I have not been in the neighbourhood of Edinburgh for several years, and have had no chance of seeing you.' Anderson (later Sir Robert Anderson) who recalled his work for the Commission in his published memoirs remarked that 'Airy and Piazzi Smyth could never hit it off.'

On returning from France at the end of June Piazzi Smyth was upset to discover that Airy had visited the observatory in his absence, though the Commission meeting was due only in the middle of July. He reacted overhastily by complaining to the Home Office of the unscheduled visit; but there was a simple explanation. Airy had merely paid a call on his way north for a month's tour of Scotland and the Orkneys. The actual meeting coincided with his return journey.

The Commission met in Edinburgh for four days, from 19–22 July. The members inspected the observatory and the official house on the first day, and on the second day interviewed Piazzi Smyth in detail. The interview, recorded verbatim, reads rather like a cross-examination, with plain speaking on both sides. Piazzi Smyth was glad of the opportunity to state his outspoken views on the needs of the observatory, on the shortcomings of the government's method of financing and administering it, on the timewasting chore of making it keep the Scottish meteorological records and on what he had come to see as the disadvantages of the observatory's university link. In a written communication he also stressed the importance of having the financial means to allow 'travel about the earth's surface' to make actual observations on sea or land under special conditions; he pointed out that in a small way and at his own private expense he had made seven 'more or less scientific journeys'. Scientific travel, he said, had been a feature of his tenure and 'it should not be allowed to fall into desuetude for want of proper funds'. Despite the critical tone of

A view from Calton Hill over the smoky city of Edinburgh with the Dugald Stewart Monument in the foreground and, in the distance to the right, the smoke from the one o'clock time gun on the castle. This interesting photograph was taken by Piazzi Smyth in 1876 (RSE).

their questions the Commissioners were in fact extremely sympathetic, and Piazzi Smyth could not have wished for a better Chairman than Lord Lindsay, his astronomical neighbour at Dun Echt, a fellow spectroscopist and auroral observer. The Commission finished their work and produced a draft report on the fourth day. The final *Report*[3], which was in fact written by Anderson, was published in November, and in the following July at the request of Duncan Maclaren MP the Report was published as a Parliamentary Blue Book. Anderson sent an advance copy privately to Piazzi Smyth, writing: 'It will, I trust, justify the assurance I gave Mrs Smyth that the recommendations of the Commission are a mean between her hopes and her fears!'

The *Report* acknowledged the lack of support which the observatory had received as a government institution. It recommended an increase of almost a hundred per cent in the annual budget and deplored the manner in which Piazzi Smyth had been obliged to keep the observatory running out of his own pocket. It declared that the observatory was understaffed and underpaid. It recorded the unsuitability of Calton Hill as a site for an observatory and recommended that at some stage in the future the observatory should be transferred to a more suitable position outside the town, with a proper residence for the Director. It did not advise renewing the faulty meridian instruments on the existing site; the Commission suggested that the Edinburgh Observatory should concentrate on physical rather than positional astronomy. 'Owing to the fact of its possessing a reflector of such a large aperture and short focus, work might be carried on at this Observatory, which could not be at present done elsewhere. In fact, if we might be permitted to do so, we would suggest that should Her Majesty's Government ever contemplate establishing what has been called a "Physical Observatory", they would have here the nucleus of such an establishment, the equatorial being eminently fitted for physical observation of the heavenly bodies, as adverted to by Her Majesty's Astronomer. We consider that such useful work is to be done in this branch of astronomy that we do not recommend that a new meridian instrument should be supplied to the Observatory—and that, irrespective of the expense, which would be considerable. And we suggest that the force of the staff should be directed to the furtherance of our knowledge of astronomical physics.'

On the university debate, the Commission could not accept Piazzi Smyth's views. They acknowledged that the rights of the university to call upon him to lecture 'should be held in abeyance during Professor Piazzi Smyth's tenure of the Chair, but that future occupants of it should be expected and required to discharge its duties'. Regarding the equatorial telescope Piazzi Smyth gave evidence of the 'series of contretemps' which made him despair of ever having an instrument of the standard he had hoped. The Commission concluded that the telescope was unfinished and unusable and left the detailed discussion of it to a later meeting.

That meeting took place three years later, in 1879, when Ralph Copeland, David Gill's successor as Director of Lord Lindsay's observatory at Dun Echt, was given the unenviable task of inspecting the telescope. Fortunately, Copeland was a man of considerable tact who got on well with Piazzi Smyth. He assembled all the correspondence, amounting to hundreds of pages, between Piazzi Smyth, Grubb and the Office of Works, and succeeded in producing a report on the whole complicated story. Some recommendations were made to the Office of Works for alterations but they were never completed, and the telescope was virtually abandoned.

It is noteworthy that the Commission saw the Royal Observatory Edinburgh as a future centre for astrophysical research and not as an imitation of Greenwich and national observatories generally with their duties to perform positional astronomy. It is interesting also that the link with the university was confirmed. In these two respects the foundations were laid by the Commission for the future character of the observatory. However, it did not allude at all to Piazzi Smyth's most cherished ambition, a mountain observatory for peripatetic astronomy. Piazzi Smyth approved of the *Report* but he was not destined to profit to any great extent from its constructive recommendations. As far as Piazzi Smyth was concerned Anderson's description of the *Report* being 'a mean between hopes and fears' was fairly accurate.

The *Scotsman*, Scotland's national daily newspaper published in Edinburgh, carried an account of the Commission's findings, together with a lengthy and highly sarcastic editoriai article. In it Piazzi Smyth's forthright evidence was recounted, interspersed with humorous comments. The very real problems of the observatory were not mentioned and there was complete disapproval of the Commission's positive suggestions. 'The Commission have entirely forfeited any claim to be recognised as authorities in moral questions by promulgating the paradox that the Professor of Astronomy's 30 years' non-performance of duty invests him with a prescriptive right to neglect it forever. If, however, we are to have a new astronomical Endowment, care should be taken that the nation gets some real astronomical work done for its money.' Tait denounced the article as 'scurrilous' and when Piazzi Smyth next visited the rooms of the Royal Society of Edinburgh he was told that 'there was only one opinion about it, that it went quite beyond the bound of both truth and politeness'. Piazzi Smyth's only comment was a note of the end of volume 14 of the *Edinburgh Astronomical Observations* which said: 'Copies of the Report were circulated to various parties, and amongst them, it is believed, to some of the Edinburgh daily newspapers, which descanted upon it accordingly.'

Chapter 12

Two Expeditions to the High Sun in Lisbon

Having acquired and tested his new powerful spectroscopic equipment Piazzi Smyth was ready to use it for the separation of true solar from telluric absorption lines in the Sun's spectrum. At the same time he hoped to be able to refine Ångström's observations of 1868 and to extend them further into the far-red end of the spectrum. In this little-studied region of the spectrum David Brewster, who had over the years made a host of contributions to spectroscopy, had discovered absorption features to the red of Fraunhofer's absorption feature A. When Fraunhofer originally observed the spectrum of the Sun he labelled the main features by letters of the alphabet beginning with the letter A at the red end and finishing with H—the double lines later called H and K—at the violet. In their map of the solar spectrum published in 1860 Brewster and his collaborator John H Gladstone, a physical chemist and later the successor of Faraday at the Royal Institution in London, had added the new features, named X, Y and Z, which, however, had not been satisfactorily confirmed by any other observer.

A solar spectroscopic programme with a similar aim had been carried out in 1875 by J N Hennessey of the Indian Trigonometric Survey. Army personnel in far-flung parts of the Empire, well-trained in the use of scientific instruments, were often employed on astronomical missions in such work as the observation of eclipses or, like Hennessey, of transits of Venus across the solar disc. Hennessey had made his observations at a site in the Himalayas at an altitude of 7100 feet with a three-prism compound spectroscope specially constructed for him with the support of the Royal Society by the Dublin firm of Grubb. Piazzi Smyth, however, was disappointed by Hennessey's spectral map, published in the *Transactions* of the Royal

Society, and was certain that he was able to do better with his own equipment. For this he needed a high sun and—to eliminate the obtrusive water vapour features—a dry climate. Tenerife would naturally have been an ideal site for such work but Piazzi Smyth was anxious to start as soon as possible and a return to Tenerife would have involved delays and financial problems. Portugal appeared to him a reasonable alternative where he could expect to see the Sun at altitudes reaching more than 70 degrees in summer and thus gain a good contrast between the solar spectra at high and low altitudes. The capital Lisbon had the advantage that it was not too far from home and could be easily reached by one of the South American steamers which regularly called there. Last not least, Piazzi Smyth could be assured of a warm reception by the Portuguese Astronomer Royal, Frederico Augusto Oom, whom he already knew and who was prepared to put the resources of the Royal Observatory at Piazzi Smyth's disposal. The fine observatory buildings in Lisbon, 'like Pulkova perfectionised', were still under construction at considerable expense and Piazzi Smyth once again compared sadly the support which continental observatories enjoyed with the meagre resources of his own establishment. As usual he had to bear the cost of the Lisbon expedition out of his own pocket, but he was able to persuade the Pacific Steam Navigation Company to carry his very cumbersome equipment free of charge.

The Smyths set off in June 1877 with seven packages of scientific apparatus travelling by train from Edinburgh to Liverpool where they boarded the 'grand and well-conducted' SS *Aconcagua* which was bound for Valparaiso with a crew of 100 sailors and 180 passengers. They had the company of their friends the Sidebothams of Manchester who were travelling to Portugal on holiday. They were in high spirits. Piazzi recorded scenes on the journey in his sketchbook and made regular drawings of the rainband spectrum while Jessica carefully maintained her meteorological diary. This first expedition to Lisbon which lasted a month was to be of an exploratory character, to be followed a year later by a second expedition of several months' duration. As a site for the observational work Piazzi Smyth first chose the hills of Cintra close to Lisbon, a well-known beauty spot celebrated by the poet Byron. However, after a fortnight there and disappointing sky conditions he moved to the grounds of the Royal Observatory in the Ajuda Park where he set up his equipment for the second time.

The equipment consisted of a manually operated heliostat whose glass mirror he had silvered himself, Salleron's double 90-degree heavy glass prisms compensated by crown glass prisms, and two-inch collimator and telescope lenses with focal lengths of 17 inches. This

Characters on board ship on the voyage to Lisbon from Piazzi Smyth's diary (RSE).

was a combination which gave a dispersion of 14 degrees between 4000 and 6500 angstroms. In all his solar work Piazzi Smyth also used an additional prism in front of the spectroscope slit, an arrangement devised to avoid the problems of coloured glass filters which were then the usual means of reducing stray light in the instrument. Unfortunately, such glasses also absorbed a great deal of light and had uneven colour transmissions, often with more than one transmission band. The whole spectroscope could be rotated and the viewing telescope was mounted on a screw which was calibrated against wavelength. Piazzi Smyth took immense care with the adjustment of this instrument. He used the prisms at minimum deviation for each wavelength range to be looked at, which meant making readjustments several times in the course of observations and using new screw calibrations every time. Since several thousand numerical calculations went into these dispersion tables the work was very laborious, but the effect was well repaid by the high precision of the measurements which put Piazzi Smyth's spectroscopic work in the top category for definition and resolution. The anterior or auxiliary prism system, which became known as the 'spectroscopical illumination pointer', was Piazzi Smyth's own special design and was later adopted by his successor in the spectroscope of Edinburgh's new Royal Observatory. The reference scale in Piazzi Smyth's spectroscope was illuminated by a light shone through a small direct-vision spectroscope so that its colour could be matched to the part of the spectrum under analysis, allowing the observer to see both the spectrum and the scale in sharp focus simultaneously. This device was supplied by Adam Hilger who made the scale of polished iridium, a metal as bright as silver which did not tarnish—though Hilger 'nearly fainted at the price'!

Piazzi Smyth tested the effects of high and low sun by observing the well-known red atmospheric Fraunhofer band B at 6900 angstroms and the strong solar lines C (Hα) at 6563 angstroms and D (sodium) at 5893 angstroms. He then observed, at high and low sun, the atmospheric band A at 7593 angstroms, the farthest to the red recorded by Ångström, and the range of Brewster's far-red features. He was able to observe to a wavelength limit of well beyond 8200 angstroms which testifies to his astonishing powers of vision, a gift he shared with Brewster, who used to prepare his 'wonderfully powerful eye' for far-red observations by bathing it in water 'to cleanse the fluid which lubricates the cornea'. Piazzi Smyth confirmed the existence of Brewster's features and was able to conclude that X and Y were solar in origin while the faint Z was telluric. When Piazzi Smyth returned some years later to a discussion of the problem he demonstrated that Y was the strong solar sodium feature at 8190 angstroms. The atmospheric feature Z turned out to be the water vapour band at about 7960 angstroms and the feature X may well have been the lines of Calcium II at 8542 angstroms. In the cramped far-red region of the spectrum as seen in prismatic instruments there was, of course, considerable uncertainty as to the exact wavelengths of Brewster's groups which, in fact, were not included in later standard maps of the solar spectrum.

Piazzi Smyth's visual investigation of the far-red region of the solar spectrum was of only limited value. This was because of the advances made in the production of photographic emulsions sensitive to red light. William Abney, then a captain and an instructor in chemistry and photography at the Chatham School of Military Engineering, developed a method of making red-sensitive emulsions and applied it to infrared solar spectroscopy. The outcome was the publication in 1879 of the first detailed infrared spectrum of the Sun reaching out to 10 750 angstroms. Abney's photography was soon followed by E Becquerel's method of tracing the infrared spectrum by phosphorescence and by Samuel F Langley's infrared bolometer which he used from 1880 onwards and with which, working at 12 000 feet above sea level on Mount Whitney in California, he recorded the infrared solar spectrum down to a wavelength of two microns. Piazzi Smyth naturally took a close interest in the work of Abney and Langley, with both of whom he corresponded. He also communicated his results to Henry Draper whose father J W Draper had been the first as far back as 1843 to record photographically the infrared spectrum of the Sun. Draper's observations corresponded to considerably longer wavelengths than the visually recorded features, and, as Piazzi Smyth pointed out in the *Edinburgh Astronomical Observations*, this gave 'the veteran scientist of New York a vast priority over every other worker in the field'. Henry Draper responded: 'My father was much gratified by the kindness you have shown in defending the absent.'

In the normal visible part of the spectrum between Fraunhofer's B and D, that is, between the red and yellow, Piazzi Smyth managed to clarify, by the superior quality of his observations, some of the confusion which existed between the findings of different observers, and to observe several new very fine lines. In this region he recorded 2000 lines as against Ångström's 1400. It was certainly true that the Royal Society Himalayan spectra observed by Hennessey were poor. Piazzi Smyth drew attention to this in an account which he published in the *Edinburgh Observations* where he also criticised, rather unnecessarily, Brewster and Gladstone's early map of 1860. It is possible that Piazzi Smyth was trying to impress his Board of Visitors at a time of great difficulty for himself and the observatory.

When the Smyths returned to Lisbon in 1878 to spend three

summer months on their further solar observations they took with them a variety of more dispersive prism arrangements for the spectroscope. Again they were warmly received by the 'fine, friendly and downright' Senhor Oom who dealt with the customs officers and arranged for the transport of their two cartloads of equipment to their rented apartments half a mile beyond the city boundaries in the direction of Cintra. They were delighted to find at the quayside their helper Francisco, a local Lisboan who had travelled in America and could speak English well. Their apartments consisted of four rooms in a house belonging to an English clergyman's widow which was formerly a school but was now given over to the cultivation of silkworms. A sitting-room looked over a lovely garden with a view of the great aqueduct. In another room Piazzi Smyth set up his temporary observatory, and with his larger spectroscope he could now quite properly claim that he was able to 'contemplate the scientific glory and mysterious physical complexity of the light of an almost zenith sun with a degree of instrumental power not yet possessed by many persons in the world'. Since the instrument-room did not face due south Piazzi and Jessica had to commence work at 5 AM in order to finish before the Sun went out of view behind some buildings in the west. The work did not prevent them, however, from seeing the sights, viewing gardens and visiting English residents in the afternoon and evening in the company of the Sidebothams who were again in Lisbon on holiday. With the temperature sometimes above 90 degrees in the shade it was an exhausting time particularly for Jessica who looked after the heliostat and occasionally felt the strain; the diaries read 'Poor Jessie ill', but next day 'up again at the usual time'.

On this second visit the Smyths concentrated on the Fraunhofer B feature at 6900 angstroms, a band due to molecular oxygen in the atmosphere, though its identification had to wait until 1885. The band had not been resolved by earlier observers; the maps of Kirchhoff and Ångström showed only broad smudges where there were double absorption lines. Piazzi Smyth was amazed at the detail which he could see with his new spectroscope, and when he asked Jessica to confirm his startling observation she exclaimed—so Piazzi reports in a paper in the *Monthly Notices of the Royal Astronomical Society*—'Oh! the beautiful double lines!' Piazzi Smyth's description of the appearance of the resolved B band is hard to beat: 'The almost proverbial beauty of the great B line must consist in the rhythmical arrangement of the powerful lines forming the preliminary band to B and its attached band of finer close set lines or linelets. Forcible strong lines the former are, clean edged, well defined, no one of them exactly like the other either in thickness or depth of colour or distance from its neighbour on either side; and yet the whole forming a harmonious

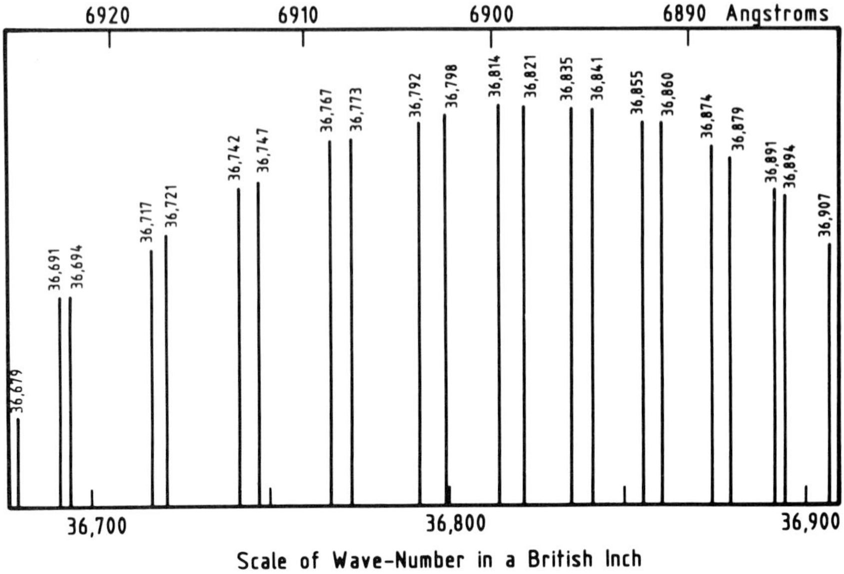

'The beautiful double lines'. The spectrum of the solar B band resolved into a series of double lines by Piazzi Smyth in Lisbon (from MNRAS vol. 39)

group, from which not one element could be taken away, and to which not one could be added, without introducing a discord and spoiling the entire system.' The band was to him an example of 'science and art combined'. Piazzi Smyth's resolution of the B band into individual pairs of lines was indeed a triumph for his prismatic spectroscope. In the same year both the A and B bands were resolved by Langley using not a prism but a Rutherfurd grating in his spectroscope. Secchi, in the second edition of his famous book *Le Soleil*, which was published in 1875 shortly before his death, had commented: 'Certain bands which in ordinary instruments seem to be stumped (*estompées*) consist in reality of numerous perfectly distinct lines, as is seen in a spectroscope possessing great dispersive power; but some of them are really massed towards the edge, and there it is impossible to separate them, however powerful the instrument employed. We may cite as an example the lines of the group B.' The work of Langley and Piazzi Smyth, the first to resolve the B band, has been overlooked in historical accounts.

In the Lisbon work Piazzi Smyth introduced for the first time his method of recording the strengths of solar absorption lines in terms of line width and line depth or blackness. Each spectrum line was represented in his maps by a line of appropriate width and height.

This new representation of solar lines by both width and depth was in fact a visual estimate of equivalent width, the first ever attempt, as pointed out by Kayser in his *Handbuch*, to record photometrically the profiles of solar absorption lines. Piazzi Smyth also converted this information on line intensities to numbers on a scale of nought to ten with finer subdivisions among faint lines, a system which was to be so effectively used by Rowland 20 years later in his compilation of wavelengths and line intensities in the solar spectrum. This method of recording the widths and depths of lines in the case of a molecular band was a striking illustration of the intensity pattern in the band, though at that time there was as yet no inkling of the origin of such patterns. Piazzi Smyth concluded that the harmony and symmetry 'show it to be dependent on one element or elementary combination, and not the chance coming together of stray lines from all sorts of alien elements scattered through the rest of the spectrum'.

At the violet end of the spectrum Piazzi Smyth's Lisbon spectra were also superior to Ångström's, the reason here being that he was able to make observations of the Sun at an altitude as high as 70 degrees and to profit from the correspondingly greater transparency of sky in the violet. The aspect of the solar spectrum seen under these very favourable conditions which most struck Piazzi Smyth was the increase in the number of lines which became visible towards the violet end of the spectrum. He used a vivid analogy in the paper he read to the Royal Society of Edinburgh: 'The continual increase in the number of the lines almost reminded one, beginning at the red end of the spectrum, of the few and far between individuals you see by the railroad side when travelling through wild hills and Caledonian moors; and then as you go on through the spectrum colours, the lines increase, just as the human figures do as you approach the house-covered suburbs of London.' He predicted that if one could observe further into the ultraviolet the ranks of lines would compare with the crowds in the centre of the city.

Piazzi Smyth made a brave attempt to interpret the increase in the number density of lines towards the violet in terms of the temperature of the Sun. It was the well-known experience of laboratory spectroscopists that, as long as their chemical investigations were confined to the temperature of a flame, the spectrum lines observed were chiefly in the red, while at the higher temperature of an electric spark there was an immediate increase in the number of lines in the blue and green. Piazzi Smyth collected from various sources all the known laboratory spectrum lines and graded them on a scale of increasing source temperature from 'freezing temperature' (the telluric lines in the solar spectrum) through the temperatures of flames and vacuum tubes to that of powerful electrical induction sparks. It was a scale of what Norman Lockyer was later to call 'heat levels'. The frequency of lines by wavenumber seemed to show a general shift towards the violet as temperature increased. The vastly greater number of lines observed in the violet part of the solar spectrum appeared to indicate an even higher temperature than that of electric sparks, but since the temperature of sparks was unknown it was impossible to estimate what that higher solar temperature might be. While Piazzi Smyth's paper was passing through the press, Langley's first bolometric measurements of solar radiation appeared in print. Comparing the Sun's radiation with that of a steel furnace Langley had arrived at a figure of 300 000 degrees Fahrenheit for the minimum temperature of the solar surface! There were enormous variations in the values suggested by various workers for the Sun's temperature. The problem was not just an observational one: it partly arose from ignorance about the radiation laws. Langley's experimental method of operation was to prove entirely right and his perfected bolometric measurements, published 15 years later, could be interpreted correctly according to the Stefan–Boltzmann law of radiation in terms of a temperature of the Sun's visible layers of about 6000 degrees Centigrade.

When Langley was preparing in 1881 to ascend to Mount Whitney in the Sierra Nevada in order to record the solar radiation and also to study the infrared solar spectrum with his bolometer he wrote to Piazzi Smyth on the subject of his Tenerife experiences. Langley had read the book *Teneriffe Experiment* and asked for a copy of Piazzi Smyth's scientific *Report* to the Admiralty, saying 'I have seen it lately in Boston, 600 miles away!, and have copied all I could referring to my subject in the brief opportunity I had of access to it.' Langley was particularly interested in the black-bulb thermometer which Piazzi Smyth had used on Tenerife to measure solar radiation and thought that this part of the Tenerife experiment ought to have attracted more attention. He was equally interested in Piazzi Smyth's early Tenerife work on telluric lines and he mentioned his own newly invented method of observing the rotation of the Sun by simultaneously projecting the light of its two edges on to the upper and lower parts of the slit of the spectroscope and looking for Doppler displacements in the lines of their spectra. This was a method which, obviously, could also be used to discriminate between lines of telluric and solar origin and was one which was widely used in the years that followed.

The Lisbon observations, and particularly the resolution of the solar B band, demonstrated without any doubt their superiority over earlier records. This improvement was due as much to Piazzi Smyth's 'green fingers' which could coax the maximum performance from an instrument by painstaking adjustment, as to the excellent optics in his spectroscopes chosen from the best Paris and English opticians. In recognition of the solar spectroscopic work of the Lisbon expeditions Piazzi Smyth was awarded the Brisbane Prize of the Royal Society of Edinburgh in 1878, his second major prize from the Society, the first being the Keith Prize for metrology of the Great Pyramid.

Piazzi Smyth's experiences in improving his solar spectroscope had led him to realise the faults of coloured glasses as a means of reducing glare in his instrument. In a search for better filters he decided to investigate the colour-transmitting properties of various liquid dyes as well as of available coloured glasses. He was also inspired by the need, felt by astronomers, for good monochromatic filters which would, for example, allow the red solar prominences to be seen without a spectroscope. (He did not realise that such a filter needs a band-pass so narrow that it was impracticable until Lyot invented the interference filter in 1931.) The absorption spectra of light-transmitting substances are responsible for their colours, and Piazzi Smyth was able to study these by sketching the spectrum of white light viewed through glass or liquid filters. Coloured media could be defined as monochroic, dichroic or polychroic depending on whether they transmitted one, two or more wavelength bands. Piazzi Smyth studied a large range of substances and measured the wavelengths of their transmission bands and estimated their percentage peak transmission. His useful absorption spectra of about thirty monochroic, dichroic and polychroic substances were published in the *Transactions of the Royal Society of Edinburgh*. The substances were not confined to coloured glasses and dyes but included port, claret and wines of various origins!

Piazzi Smyth also found himself attracted to a phenomenon which had been noted by Brewster among others that some dichroic substances appear to change their colour if their strength is changed, with one of the two transmission bands getting 'an advantage over the other'. He also thought that he could notice what he called 'locomotion' in the colour bands, the peak wavelength of the transmitted band shifting with the intensity of the light source. These effects were of course physiological and the whole interesting subject had already been studied by Helmholtz and explained in 1868 in one of his popular lectures which were translated into English in 1873, though Piazzi Smyth was obviously unaware of Helmholtz's work.

During the second Lisbon tour Piazzi put his coloured media to another use by painting a whole series of what he called 'dichroic' pictures. These were pictures of scenes and objects seen through dichroic filters. Piazzi Smyth had a selection of commerical 'Judson's dyes' which he put in flat glass bottles to form colour filters; then, using one filter after another, he painted what he saw through each. He studied the manner in which tones as well as colours in a picture change when recorded through different filters. The results were sometimes surprising on account of the two transmission wavebands; objects which seemed to have the same colour when looked at normally could be of markedly different hue when seen through a dichroic liquid. It seemed to him that his observations were relevant to the problem of colour blindness. This was a subject which he had heard enthusiastically discussed in his early days in Edinburgh by George Wilson, the first Director of what became the Royal Scottish Museum and Edinburgh University's first and only Professor of Technology, who had regarded his researches into this phenomenon, for which he received a prize from the Royal Scottish Society of Arts, as one of his greatest scientific achievements.[1] Wilson had demonstrated that the well-known defect which shows as an inability to distinguish the colours red and green is fairly common in the population and argued forcibly against the choice of these two colours for railway and marine signals. To the normal eye all colours could be produced from a combination of three primary colours, green, red and blue (or violet); in colour blindness two of these colours appear as one. The normal eye could therefore be described as trichroic and the colour-blind eye as dichroic. By choosing a filter which had only two transmission bands, red and violet, Piazzi Smyth could simulate the effect of colour blindness. Some of the dichroic scenes which he painted were very striking, with red mountains and vegetation and various shades of pinks in the landscape, while the sky remained vividly blue; he thought them indeed much more beautiful without the yellow-green (what he called 'citron'), the third primary colour. Piazzi Smyth recalled 'a late and most esteemed friend in another part of the world'—his former colleague William Mann at the Cape who was colour-blind but was also an artist—who 'though he could recognise an immense variety of tints, including dark shades, in Nature; and could aesthetically enjoy their brighter beauties and also their more sombre chiaroscuro effects and deeper shadows', yet, could not see 'a particle of colour difference' between a green leaf and the red flower of the pomegranate. At the end of a paper on colour read to the Royal Society of Edinburgh Piazzi Smyth discussed the aesthetic aspect of his experiments. 'Can it ever be desirable, in the beautiful, the exalted Fine Arts to see anything with an eye that makes green grass red?' he asked. He answered 'yes'; by taking a dichroic view of the landscape an artist could dispose of the crude colour green and reduce the overpowering yellow of which the normal eye sees too much. He gave the example of the Dutch and Flemish landscape painters 'striving in their day after noble ideals rather than mere realistic transcripts who eschewed green altogether and employed "warm brown" instead, giving the grass lawn, in the words of an unnamed author, the "exquisitely melodious tint of an old Cremona violin"'. He also discussed a landscape by Titian in the Scottish National Gallery which showed trees and grass rendered in the same 'ancient violin hue' and Turner's masterpieces of sunsets.

Piazzi Smyth wrote a most entertaining account of his Lisbon experiences in the *Astronomical Register* (a magazine which reported meetings of astronomical societies and printed letters and articles of general interest) which showed that he had lost none of the zest for life which he had displayed in the travel books of his younger days.

Chapter 13

Madeira, the Ocean Flower

While the results of the two Lisbon expeditions were being published Piazzi Smyth was already planning one or two further expeditions. The first was to be to Tenerife where he intended to improve his Lisbon observations of the spectrum of the high Sun by using a diffraction grating instead of prisms in his spectroscope. In a second expedition, to take place a year or two later, he hoped to be able to observe the spectrum of the Aurora to advantage from an arctic site during sunspot maximum. With his laboratory spectroscopy proceeding very successfully he thought that he might indeed be able to discover the origin of the lines in the auroral spectrum. He placed an order with L M Rutherfurd of New York for a grating of 1.6 inches square with 17 000 lines to the inch which was ruled by D C Chapman and tested personally and described as first class by Rutherfurd himself. With this grating, which arrived in Edinburgh in August 1880, he had a spectroscope which was second to none.

However, before he could embark on another expedition Piazzi Smyth had to cope with some personal problems. There was first of all the deteriorating health of his wife, his faithful and ever-present assistant, who suffered from rheumatism and found it painful to walk. To help her condition the Smyths visited in the spring of 1880 the hydropathic health resort of Malvern, the first of many prolonged visits to that and other similar resorts in England. Piazzi himself, otherwise so robust, was at the same time getting increasingly deaf. He took to carrying a slate about with him on which people could write, and in 1880 he acquired a large ear trumpet which from then onwards was taken on all his travels.

The Smyths combined their visit to Malvern with a call at nearby Hereford on the widow of the recently deceased Henry Cooper Key who had been particularly helpful to Piazzi Smyth in his difficulties with his equatorial telescope. While in Hereford he had the opportunity of meeting G W With who lived nearby. With, a successful maker of silver-on-glass mirrors of up to 18-inches diameter, told Piazzi Smyth that a 24-inch short-focus mirror of high quality ought to be easily attainable, and that Dr A A Common of Ealing, London, was now 'the man for large apertures'. Common was then engaged in the making of a 36-inch reflector for Edward Crossley's private observatory at Bemersyde in Yorkshire, an instrument which Crossley later presented to the Lick Observatory in the United States where it became the first large reflector in the country.

In the difficulties of those years Piazzi Smyth must surely have felt the absence of one of his closest friends, David Gill, who in February 1879 was appointed to the post of Her Majesty's Astronomer and Director of the Royal Observatory at the Cape. The two friends had now to converse by letter, and this they certainly did, Piazzi Smyth writing a letter of ten foolscap pages recalling his own years at the Cape and Gill replying 'as a very old and true friend' with another long letter in which he gave an account of his adventures in observatories on the Continent and of his new responsibilities at the Cape.

The year 1880 marked the end of an era in the history of the observatory on Calton Hill in that on the last day of that year Alexander Wallace retired from his post as First Assistant. Then 77 years old he had served the observatory with unwavering devotion for 47 years, 35 of them under Piazzi Smyth. He had carried the burden of the tiresome and soul-destroying meridian observations with the transit instrument and their reduction from Thomas Henderson's arrival until Piazzi Smyth's solar motion programme came to an end. He died two years later. Piazzi Smyth was very fortunate with Wallace's successor, Thomas Heath, a graduate with a first-class honours degree in mathematics from Trinity College, Dublin, who was promoted from the post of second assistant and moved into the little residence on Calton Hill. Heath took a great interest in Piazzi Smyth's spectroscopic work and willingly undertook the tedious tasks of computing wavelength tables and preparing diagrams and spectral maps. He remained a member of the staff of the observatory until his retirement in 1905 and was actively involved under Piazzi Smyth's successor in the major move of the observatory to its new site on Blackford Hill.

In planning his expedition to Tenerife for the summer of 1881 Piazzi Smyth applied, through the Professor of Astronomy in Madrid, to the Spanish authorities in the Canary Islands for permission to observe there. The permit was granted early in that year with assurances of cooperation from local authorities on the island in the form of a 'Royal Order' communicated to the Governor. Optimistically Piazzi Smyth also requested a grant from the Spanish government and recommended that a communicating cable be laid between the mainland and Tenerife. He was informed, however, that the government 'had not enough money for necessities and while in debt can hardly afford such expensive luxuries as a Peak-side Observatory', a reply which cannot have occasioned much surprise. Meanwhile the practical question of travelling to Tenerife had to be looked into. On his first mission in 1856 Piazzi Smyth had enjoyed the advantage of a private yacht for his own exclusive use, but to reach Tenerife by commercial steamer would be difficult and expensive. No doubt he would have found a way round the problem if his wife had been in better health, but in the circumstances he had to consider an alternative site.

He decided on the island of Madeira as being further south than Lisbon or Malta and a place which could be easily reached by Cape-bound steamers at the moderate cost of £20 each way. It was an island at low latitude where the noon sun stood only 10 degrees from the zenith at the solstice and where Piazzi Smyth reckoned on finding in the vicinity of Funchal a site at 200 to 500 feet above sea-level protected from the prevailing north-east winds. As regards astronomical seeing conditions he had little to go on; a paper in the *Monthly Notices of the Royal Astronomical Society* describing observations of Mars made there in 1877 was not very useful but he hoped the site would be as good as Lisbon. Accordingly, on 10 May 1881, he and Jessica sailed from the East India docks in London on the 'splendid steamship' RMS *Dunrobin Castle* bound for Madeira, their passage partly assisted by the ever helpful R M Smith of the Observatory Board of Visitors, the same friend who had arranged their voyage to Russia more than 20 years before. Their week at sea was recorded as usual in words and sketches, as were the meteorological and rainband readings and the daily geographical positions of the ship.

Entering Funchal Bay on the evening of 17 May Piazzi Smyth was 'aghast' to see the hills of the island 'bright green from top to bottom, exhibiting a much more fertile view than the neighbourhood of Lisbon or Santa Cruz, Egypt or Cape Town so that the climate must be as we see it now, utterly cloudy'. As the ship came into harbour there were little boys diving for coins into the sea, a sign that the waters were warm and the atmosphere moist, no environment for 'the clear skies for spectroscopy'.

Jessica had to be carried ashore and up to Jones' Private Hotel where they had arranged to stay and to set up their observatory. As in Lisbon their friends the Sidebothams from Manchester were again with them. The garden around the hotel was smothered in exotic flowers and tropical plants whose names and descriptions Piazzi Smyth recorded minutely in notes and watercolours. The Jones' establishment which catered for convalescents was the height of luxury and comfort with 'meals grandly served, silver superbly polished and tablecloths snow white'. Jessica's condition, however, was so painful that she had to be assisted up and down stairs and carried in a hammock when they paid visits to the many hospitable families which made up the large British colony in Madeira. These included the Blandys, the owners of the fabulous gardens, one of whose twelve children was the wife of Sir William Thomson, the physicist.

Piazzi set up his 'heliostatic observatory' in the stone-floored former smoking-room of the hotel. The room faced south-east, so he got up at dawn each day to make the most of the hours of accessible sunshine and was already hard at work by 6.30 when he was brought his porridge. In spite of her ailment Jessie carried on bravely operating the heliostat 'with patient enthusiasm and enduring skill' during the observing sessions. Though the sky was often cloudy and the atmosphere moist, conditions could be remarkably good, and at night the Zodiacal Light was brilliant.

The spectroscope used in Madeira had its prisms replaced by the new Rutherfurd grating. The heliostat protruded out through the otherwise black-curtained window sending the solar beam into a six-inch achromatic lens of 90 feet focal length. As in the Lisbon arrangement there was an anterior prism to remove unwanted light from the solar beam. The resolution of the new grating spectroscope was a considerable advance on the former one as regarded the far-red region of the spectrum, so that the A band was easily resolved into its pattern of close pairs of lines like those recorded in the B band in Lisbon. Piazzi Smyth likened the A and B bands to the two Magellanic Clouds, and their resolution into lines to Herschel's resolving of the Clouds into individual stars.

At the time there were still many discrepancies between the records made by different observers of the more complex features of the solar spectrum, arising from the very different resolutions of their various spectroscopes. Piazzi Smyth therefore planned to produce his own independent and improved version of some of the main features of the spectrum at high sun and thus with the least possible contamination by telluric lines. He also hoped to observe the same features at low sun in order to recognise those actual telluric lines. He chose fourteen spectral regions between 4300 and 7800 angstroms, each about 50 angstroms wide and each centred on one of the main features indicated by Fraunhofer or Brewster. He was able to complete his

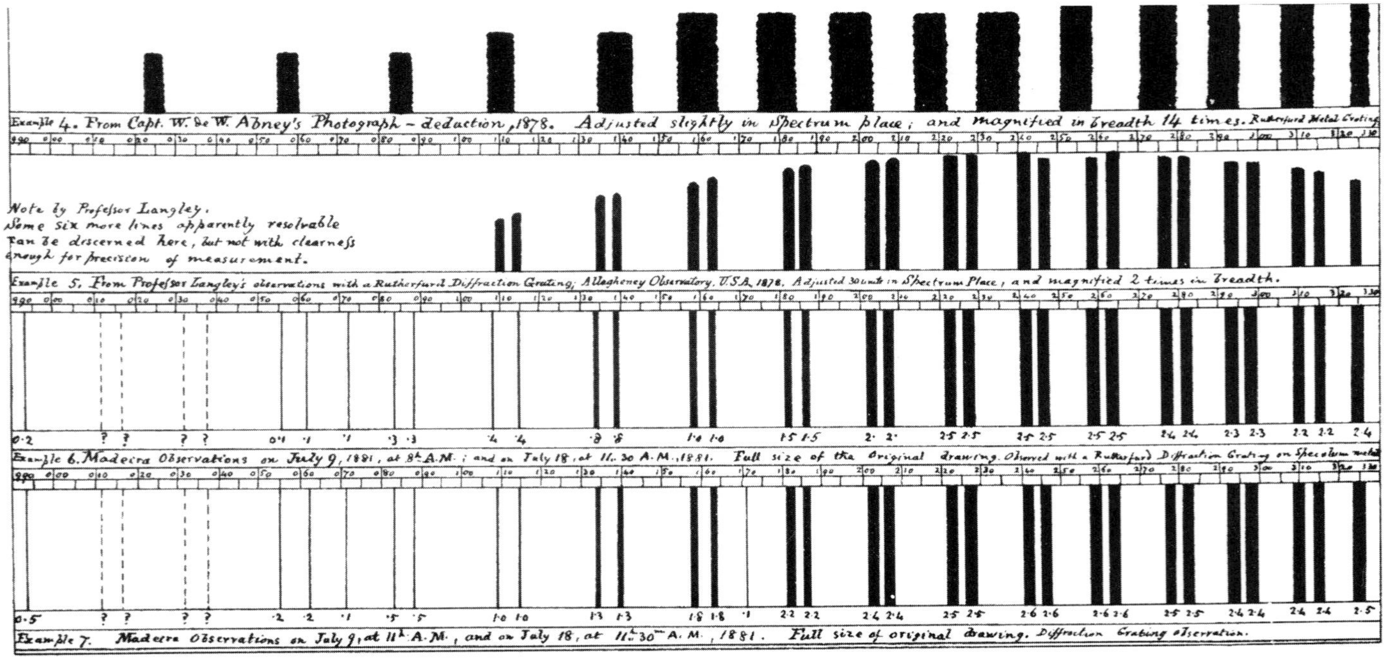

The A band of atmospheric oxygen resolved in Madeira (two lowest spectra) and compared with the photographed spectra of Abney (top) and Langley (Madeira Spectroscopic).

programme of high-sun spectroscopy but had time for only half of the low-sun spectra, those in the yellow and red regions which as he already knew included some of the strongest telluric features.

During their two-month sojourn in Madeira the Smyths had the pleasure of a morning visit from David Gill and his wife, when their Cape-bound ship called briefly. They also had the sad experience of hearing in a letter from Piazzi's brother Warington of the death of their sister Ellen after a long illness. The wife of the meteorologist Henry Toynbee, and, like the Smyths, a great lover of travel, Ellen had been particularly dear to both Piazzi and Jessica.

When the time came to return home, the Cape steamer picking up passengers for England arrived at midnight. Embarking of passengers and luggage, including the contents of the heliostatic observatory had to be done in the dark, and Jessica, unable to walk, had to entrust herself to other hands to transport her aboard. Piazzi wrote down a vivid account of this experience in the early hours of the following morning. 'After having fortunately taken leave of everyone in the hotel yesterday evening and retired to bed, we were called up at midnight with the news that our steamer was entering the Bay. Sledge and hammock having arrived they were hastily loaded up by starlight and one flickering candle, and we moved off through the darkness with all our belongings through lanes darker than dark under the shadow of the garden trees of Madeira.'

The passengers on board ship were mostly military men, some returning home wounded from the war in the Transvaal. In spite of Jessica's disability Piazzi kept a cheerful record of the return voyage. A week later they docked in London and after an overnight stay there they reached Edinburgh by the evening train, 'thankful to arrive safe after the long tour'.

The Madeira spectroscopic results were published in the form of spectral maps arranged together with earlier ones of other observers, all reduced to wavenumbers and shown for comparison on the same scale—though there seemed little purpose in going back as far as Brewster and Gladstone's outdated map of 1860. Piazzi Smyth's justification for making the comparisons was to establish for the benefit of theoreticians 'how closely the most improved map-versions hitherto in vogue and on which new theories largely rest, do really represent in certain crucial instances the exacter phenomena of the Sun's spectrum'. The superiority of Piazzi Smyth's spectra over all but one of the others was unmistakable, the exception being the new photographic prismatic spectrum of the blue published in 1881 by Vogel of the recently founded Astrophysical Observatory of Potsdam. Apart from minor differences in the case of very faint lines, the two versions confirmed each other in the regions common to both sets of

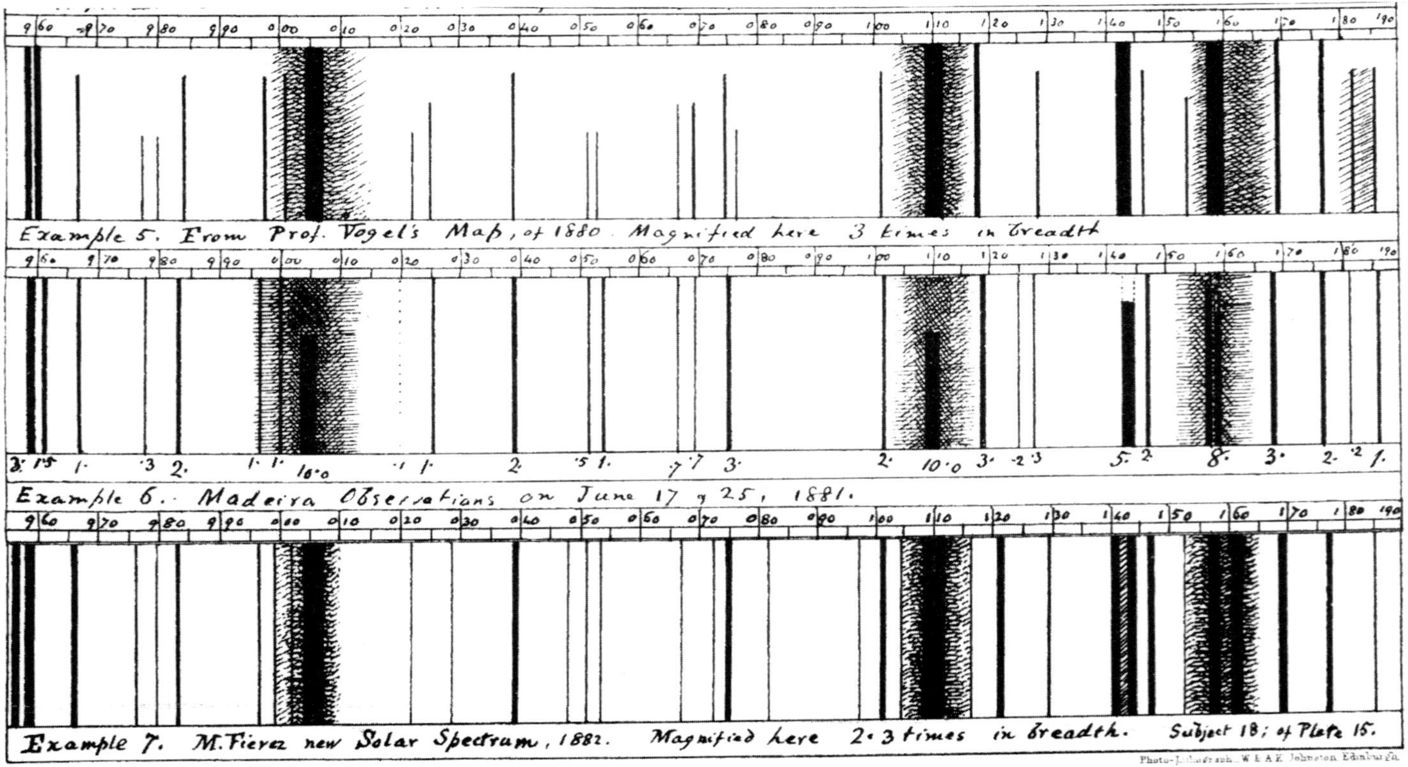

Three observations of the green triplet of magnesium in the solar spectrum by Vogel (top), Piazzi Smyth (centre) and Fievez (bottom) (CPS Madeira Spectroscopic).

spectra. Piazzi Smyth's visual observations captured almost every line recorded photographically and, in his own view, with better definition. It is indeed a remarkable tribute to Piazzi Smyth as a practical spectroscopist that he was able, with no public support and with equipment assembled by himself in his own private laboratory and at his own expense, to match in quality the work of Vogel, one of the most distinguished astronomical spectroscopists of his generation.

The high sun at noon in Madeira had provided better observing conditions than Piazzi Smyth had ever previously attained. A comparison of high and low sun seen in the moist conditions of Madeira with those seen from the dry station at Lisbon allowed him to separate, for the most part correctly, the true solar, the atmospheric dry air and the atmospheric water vapour absorption lines among the features which he investigated. He was the first to do this, though conflicting attributions were proclaimed at the 1882 meeting of the British Association by other spectroscopists. Among the atmospheric features confirmed as non-water vapour were the two bands in the red, Fraunhofer's A and B. An interesting water vapour feature identified and resolved was the group of lines in the vicinity of the sodium yellow D lines, not far from the main rainband, originally observed by Cooke in 1866. A smaller band in the red designated α was shown not to be a single feature but a mixture of telluric and solar lines; the French spectroscopist A-M Cornu chose Piazzi Smyth's observations of this group to test his own device for distinguishing solar from telluric lines. Piazzi Smyth's analysis of these telluric elements has received less than its due recognition in historical accounts including Scheiner's *Astronomical Spectroscopy* of 1890.

Piazzi Smyth's results, printed in 1882 in a special volume called *Madeira Spectroscopic*, aroused considerable interest at the time, in particular his detailed map of the Great A band which surpassed Langley's photographic record. The similarity in the structures of the A and B band caused Piazzi Smyth, and no doubt other spectroscopists as well, to surmise that the two bands had the same origin. Piazzi Smyth's guess was that they were caused by nitrogen, the most abundant gas in the terrestrial atmosphere. In fact the A and B bands were found to be due to oxygen, a common element on Earth which astronomers had been searching for in the Sun but which had so far eluded them. The identification was demonstrated by N Egeroff, a Polish spectroscopist who succeeded in observing the B band in the spectrum of an electric light observed at a distance of 10 kilometres, and also in the spectrum of oxygen seen in absorption in a tube of that gas. This proved that the B band in the solar spectrum is produced by 'cool' (molecular) oxygen in the air, though many were reluctant to accept the apparent absence in the Sun of such a basic element as

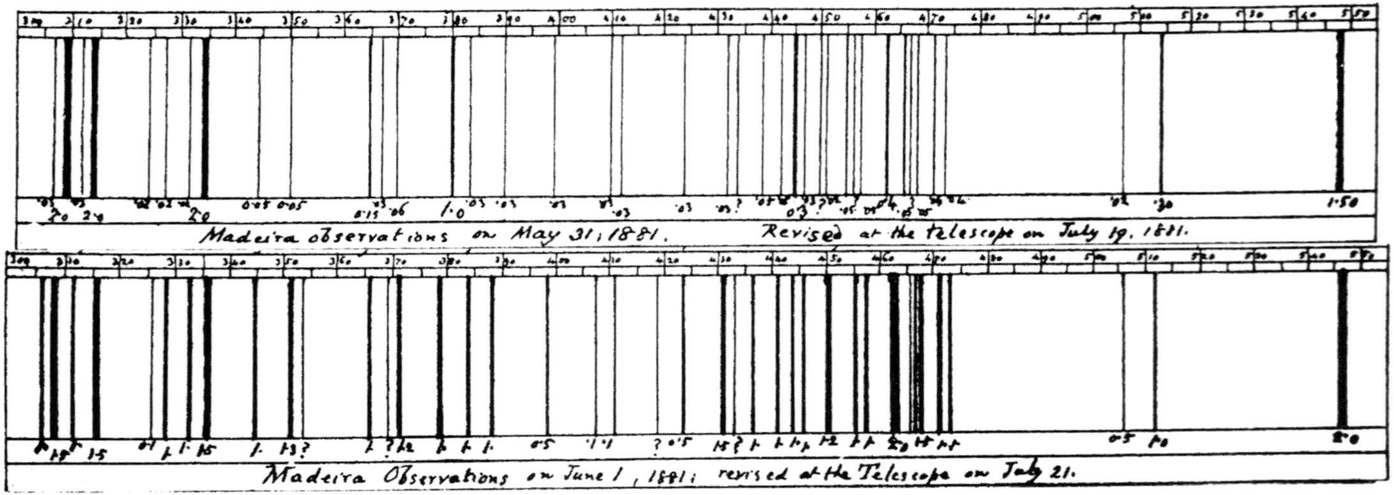

A 'band' or group of lines (spanning 40 angstroms at 6300 angstroms), resolved for the first time by Piazzi Smyth, at high and low Sun. The strengthening of some of the lines showed it to be a mixture of solar and telluric lines; the 'eye and hand' method is not without ambiguity and was not used again by Piazzi Smyth for telluric line identification (Madeira Spectroscopic).

oxygen. Piazzi Smyth was at this time also working on the laboratory spectrum of atomic oxygen, and was thus deeply involved in the puzzle of the missing oxygen in the Sun.

These considerations revived a matter which Piazzi Smyth had discussed in connection with his Lisbon work. A suggestion had been made by Abney that some of the lines in the solar spectrum might be produced by gases in the interplanetary medium. Piazzi Smyth agreed that the high-sun and low-sun method would not be able to distinguish such lines from purely solar ones. In the case of the B band, which clearly increased in strength at low sun, Piazzi Smyth had remarked that the feature was still strong at high sun in Libson so that on the basis of these observations alone he could not categorically rule out the possibility that the Sun itself, or the intervening space, could be responsible in some part for the feature. The Madeira observations confirmed the very notable growth in the A and B bands as the Sun approached the horizon, indicating that any non-telluric component in these bands would have to be small. When the French spectroscopist Thollon later published his own observations of the B band made in the Observatory of Nice he found fault, through a misunderstanding, with Piazzi Smyth's remarks. In an article in *Nature* Piazzi Smyth repeated his argument that until observations were made from high elevations there could be no certainty as to whether or not some proportion of the B band was produced in the Sun's atmosphere itself.

By this time an independent method of distinguishing between solar and telluric lines was already being used by a number of observers including Thollon himself. This was the simultaneous observation of the spectra of the two limbs of the Sun, which showed, because of the Sun's rotation, a Doppler shift in the true solar lines. The method had been introduced by Vogel as far back as 1871 but Langley and others had in the meantime applied the same principle to their own work. These observers, however, were mainly concerned with measuring the Sun's rotation using telluric lines only as stationary standards. Cornu in France began systematic observations of the entire spectrum by this method in 1883 and published his important paper on telluric lines in 1886; in 1890 Thollon's atlas of the solar spectrum, discriminating between solar and telluric lines appeared in the *Annals of the Observatory of Nice*. However, the possibility that the oxygen bands could have their origin partly in the Sun itself, as Piazzi Smyth had stressed, was not discarded. The problem was still unresolved as late as the 1890s when Janssen in his old age made his famous expedition to the top of Mont Blanc and actually set up an observing station at that inhospitable site. He failed to find any trace of solar oxygen as revealed in the telluric A and B bands. As Agnes Clerke said in her *History of Astronomy* of 1893, 'our pabulum vitae is not at present recognisable as a solar element'.

Piazzi Smyth's volume *Madeira Spectroscopic* also included an account of further experiments on colours and colour filters. He tried to find a scheme for describing all colours, whether pure or mixtures of primary colours, in spectroscopic terms. The result was a beautiful colour chart, which depicted 37 different shades of colour each shown at six different levels of brightness. His experiments were reminiscent of Clerk Maxwell's colour-top experiments; indeed Piazzi Smyth seemed to think that his work might supersede Clerk Maxwell's but in fact it threw no new light on the problem. His spectral subdivisions, however, were potentially useful for observers of stars, as a means of describing the star's colour, and Piazzi Smyth advocated the adoption of his 'colour code' for the purpose, in the manner of choosing shades from a house painter's colour chart. His father, W H Smyth, a generation earlier, had proposed a similar system of 'sidereal chromatics' for describing the colours of double stars.[1] Piazzi's scheme might have attracted more response among astronomers if it had been less elaborate; he did not appreciate that the average eye is not as sensitive to variations in light and colour as his own.

In addition to his *Madeira Spectroscopic* Piazzi Smyth also published a volume called *Madeira Meteorologic*, based on a lecture to the Royal Society of Edinburgh, which was a discursive account of the weather and of the animal and vegetable life of the island, with poetic reflections on the beauties of 'Madeira the Ocean Flower' thrown in. It also served as a vehicle for the publication of Jessica's meteorological diaries in Madeira and Lisbon.

Jessica's health remained poor, and from then onwards the Smyths' life was punctuated with visits to English spas at Harrogate, Buxton and Malvern. The expedition to Madeira was Piazzi and Jessica's last astronomical journey abroad. Piazzi's plan for an Arctic expedition to study the Aurora went no further.

In between his solar expeditions Piazzi Smyth spent his time with his Aurora spectroscope in his north-facing laboratory. The spectroscope was as versatile as it was powerful. A reference spectrum for wavelength determination could be viewed simultaneously above and below the central zone of the field of view. The instrument which had grown out of the original Aurora spectroscope was now mainly used for laboratory spectroscopy. It was particularly successful for investigations of the spectra of molecules. By systematically increasing the dispersion and resolution of his instrument Piazzi Smyth was able to see the molecular bands first as wide features with one sharp edge and fading away indefinitely on the other; then being resolved into more and more 'needle-like' lines, until with sufficient resolution no further splitting of lines occurred. 'They have reached their ultimate condition and behold how exquisite they are; how beautifully ordered in their ranks, how varied in their groups, how perfect in structure.' Piazzi Smyth with his spectroscope used to the limit of its efficiency by perfect adjustment became recognised as one the leading molecular as well as solar spectroscopists of the day. He was, in fact, specially invited by the editor of *Nature*, Lockyer, to contribute a series of articles on his spectroscopic researches in 1880.

Higher dispersion in the spectroscope raised the problem of a diminution in the brightness of the spectrum. The light sources were Plücker–Geissler vacuum tubes containing small quantities of the appropriate gas electrically excited. The light from these fine capillary tubes was rather faint and it occurred to Piazzi Smyth that the amount of illumination per unit area falling on the slit of his spectroscope could be greatly increased if the tube was viewed end-on along its cross section instead of along its length. Piazzi Smyth designed a modified form of the Plücker tube to be viewed in this way and had a large number of these tubes containing different gases constructed for him by Salleron and his successor Demichel and the London firm of Louis Casella. Though the illuminated area was as small as a pinhead the effect on the brightness of the spectrum by the end-on method was 'startling in the brilliancy of the principal lines and in the immense numbers of additional lines in almost every tube spectrum examined'. The invention of end-on illumination described in a 40-page paper earned for Piazzi Smyth the Keith Prize of the Royal Scottish Society of Arts in 1879. Piazzi Smyth was not aware at the time that the same principle of end-on illumination had been applied a few years earlier by the Belgian spectroscopist van Monckhofen, but he always drew attention to Monckhoven's prior work whenever he described his own end-on tubes.

Another problem encountered by the early spectroscopists was the presence of impurities in the tubes. There were sceptics who criticised spectroscopists for apparently accepting the suppliers' identification labels on their vacuum tubes and thus running the risk of erroneous attributions to their spectra. The experience of the older generation who found the yellow sodium lines everywhere was not forgotten. Piazzi Smyth was himself extremely thorough in dealing with this matter; by his close interest in the making of tubes and the methods of producing and purifying gases, he was able to acquire tubes of the highest quality.

Piazzi Smyth's analysis of 20 gases, some atomic some molecular, with the relatively low-dispersion 52-degree prism in his spectroscope

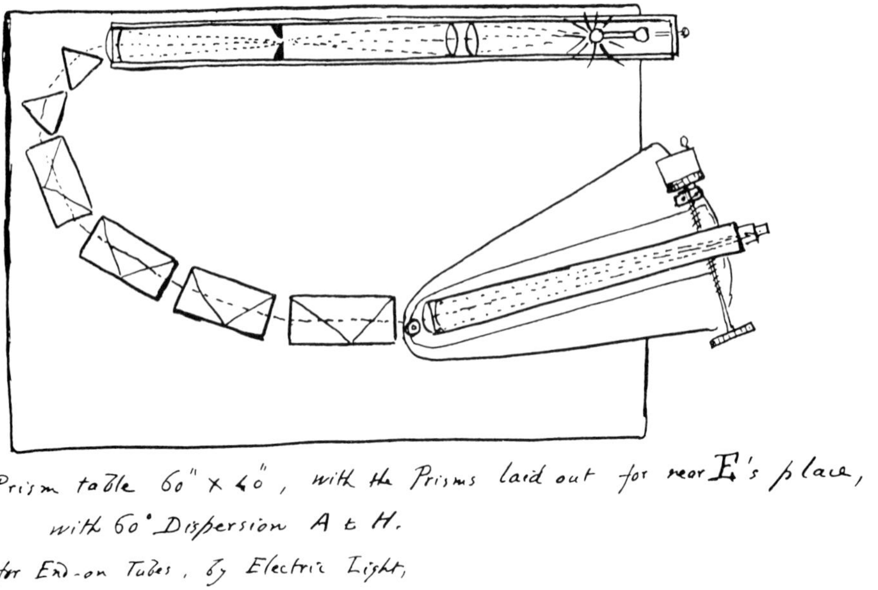

One of Piazzi Smyth's laboratory spectroscope arrangements, taken from one of his notebooks (ROE).

was published in 1880 in a paper which is as remarkable for its analysis of the effects of age and leakage on the spectra of the tubes as of the spectra themselves. It was one of Piazzi Smyth's purposes to examine the spectra of the most common chemical elements, oxygen, nitrogen and hydrogen and their compounds which in parallel he hoped to discover in the Sun. The molecular gases which he analysed spectroscopically included cyanogen (CN) and carbonic acid (CO) with their complex band spectra.

Being desirous to observe the same spectra with higher dispersion Piazzi Smyth experimented with fluid prisms and with large compound prism trains. James (later Sir James) Dewar at Cambridge lent him a hollow prism and a sample of liquid for his first experiments. Balancing the advantages of high refractive index against the many well-known disadvantages of liquids, Piazzi Smyth persuaded Adam Hilger to make for him a series of hollow glass prisms filled with bisulphide of carbon, a task which Hilger did not relish but which he performed as a favour for his admired customer. Piazzi Smyth admitted that 'the troubles poor Mr Hilger had to go through were almost overwhelming' with so many breakages of the glass blocks before a perfect hollow prism was achieved. Hilger, however, had more serious reservations: it was his opinion that it would be impossible to maintain the adjustment of a liquid-prism spectroscope for a sufficient length of time and that, eventually, the effort was better spent on improving conventional glass prisms. Many spectroscopists who had tried liquid prisms had indeed lost patience with them but Piazzi Smyth remained enthusiastic about these prisms and used them frequently for many years. Though he found them sensitive to variations in temperature of as little as one degree Fahrenheit he considered this a price worth paying for the increased resolution of individual features. Intervals of a twentieth of an angstrom could now be readily distinguished.

Using his liquid prism and his end-on illumination Piazzi Smyth re-observed his spectra and published them in 1884 in one of his most important spectroscopic papers entitled 'Micrometric measurements of gaseous spectra under high dispersion'. The title of the paper echoed a work 'Micrometric measurements' which had been written nine years earlier by the famous Swedish experts, the late Ångström and Thalen. In addition to the spectra of the 'elemental' gases hydrogen, oxygen and nitrogen, thousands of previously unresolved or only barely resolved lines in the band spectra of CO, CH and CN were analysed and some entirely new bands were discovered. All these spectra were shown on charts in which each line was depicted, according to the method which Piazzi Smyth had already used in his solar spectra, to indicate its strength.

A most important discovery in this work was that almost all the lines in the spectrum of oxygen which had previously appeared as single were in fact triplets. Piazzi Smyth resolved and measured the wavelengths and separations of six sets of these. The triplets, though widely spread through the spectrum from red to blue, showed a characteristic pattern in the relative separations and relative intensities of their three lines. This pattern is the result of the splitting of the quantised energy levels in the atom according to rules entirely undreamt of at that time. In Piazzi's own graphic words, 'the characteristics of an oxygen spectrum are so peculiar and so closely adhered to in every instance that I have been able, in some tubes swarming with impurities, to pick out an oxygen triplet as easily as one would distinguish in a crowd of civilians a soldier with cross-belts and scarlet coat'. Piazzi Smyth searched for these distinguishing signatures of oxygen in Ångström's map of the solar spectrum but could neither affirm nor refute their presence among the crowds of lines. He also noted that the ordered pattern of the green triplet in the spectrum of magnesium which he also produced in the laboratory was remarkably like the oxygen triplets. The magnesium group, called 'little b', is very conspicuous in the solar spectrum. Perhaps, he thought, the oxygen triplets were also there. His next undertaking was to include a thorough search for these oxygen lines with an improved solar spectroscope. In the event he failed to find them, but it was indeed one of the family of triplets which was to demonstrate at long last the presence of oxygen in the Sun. This was finally achieved only in 1897 when the triplet at 7775 angstroms in the far red was recorded photographically in the solar spectrum by McClean and identified as one of the oxygen group. Piazzi Smyth had actually recorded this group in his laboratory but failed to resolve it. Unfortunately he had not been able to observe it visually in the Sun on account of its extreme faintness. The other triplets are even fainter and have never been positively identified in the solar spectrum.

The discovery of the triplet groups of lines in the spectrum of oxygen was of particular interest to the young spectroscopist Arthur Schuster, then only about 30 years of age, a pupil of Clerk Maxwell's at Cambridge who later became Professor at Manchester and the country's leading spectroscopist. Schuster was at this time without laboratory facilities but had plans to work in the same field of spectroscopy as Piazzi Smyth: 'I am especially interested in the connection between the spectra of compounds and that of its elements and for this reason I shall try to map as many compound spectra as possible', he wrote to Piazzi Smyth. 'We might for instance have vibrations in a molecule due to the forces which bind atoms together and others which are due to the atoms themselves.' Schuster had observed the spectrum of oxygen and had recognised two varieties of lines—single lines produced at high temperature and others, the 'low temperature lines', which he suspected of being compound though he had failed to resolve them himself. When Piazzi Smyth succeeded in

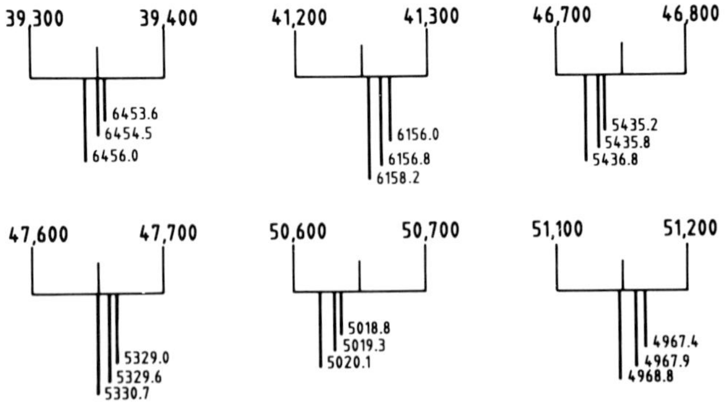

The triplet series in the spectrum of oxygen discovered by Piazzi Smyth. Piazzi Smyth's scale in inverse inches is at the top in each case; modern wavelengths in angstroms have been added (adapted from Trans. RSE vol. 32).

resolving the triplet lines Schuster wrote: 'I congratulate you on the definition of your spectroscope. I know pretty well what the different spectroscopes in this country and abroad can do, and there can be only very few indeed which could resolve these lines.' The distinction between low-temperature and high-temperature lines in this particular case was due to neutral and ionised atoms respectively, not to the molecular and atomic states.

It is a shame that Piazzi Smyth's important discovery of the oxygen triplets, like his resolution of the molecular oxygen B band in the solar spectrum, has not been adequately remembered among astronomers. In discussing the finding of oxygen in the Sun Agnes Clerke refers only to Paschen's much later analysis and makes no reference whatever to Piazzi Smyth's work or to the part he played in the search for the elusive solar oxygen.

Piazzi Smyth's analysis of the green band of CO in collaboration with A S Herschel and their discovery of what was called the 'harmonic law' is another historic spectroscopic advance which was published in the paper on 'Gaseous spectra'. Using Piazzi Smyth's data Herschel succeeded in solving the pattern of the band as made up of two exactly similar arithmetical progressions, one consisting of single strong lines and the other, displaced from the first, of fainter, closely double ones. Piazzi Smyth illustrated the progressions by drawing the two theoretical series on the same scale to show how perfectly they agreed with the observations. What Herschel had discovered and what Piazzi Smyth had illustrated was what much later became known as the branches in a rotation band and the Deslandres formula for the frequencies of individual lines in the branches. There has never been any dispute about giving the two collaborators priority in this important discovery of molecular spectroscopy. Piazzi Smyth's practice of plotting maps of spectra in wavenumbers rather than in wavelengths facilitated the solving of these laws by Herschel. It was unfortunate, however, that Piazzi Smyth had taken to giving wavenumbers in inverse British inches making no concession to other scientists who by this time had all adopted metric units. A S Herschel suggested to his friend that the dilemma could be solved by using frequencies instead of wavenumbers, a style which would be independent of the units of length used. Such units of frequency (hertz) were in fact adopted by radio astronomers from the first. It is a pity that Herschel did not insist on this usage which would have allowed spectroscopists to make practical use of the thousands of lines measured by Piazzi Smyth in the spectra of various molecules. The spectra of oxygen and CO were described at the meeting of the BAAS in 1885 when Piazzi Smyth bowed to necessity by giving the separations of his lines in metric

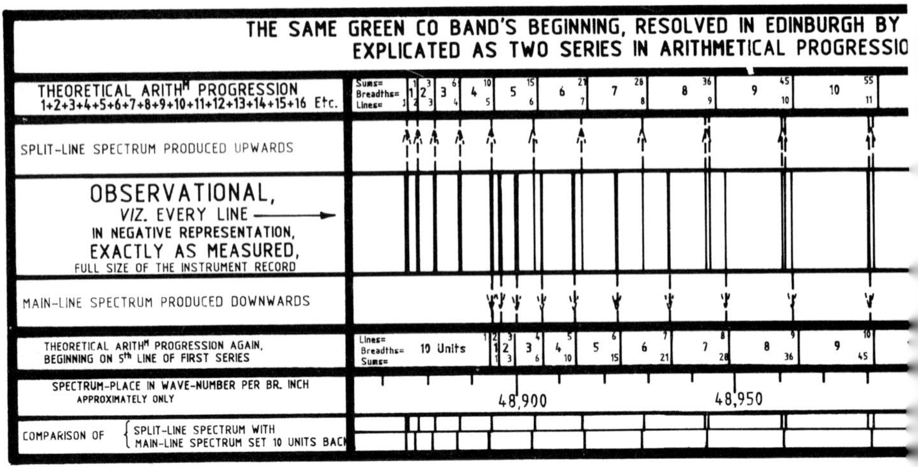

The spectrum of the CO molecule and its analysis. The arithmetic series of lines in the top row, combined with a similar series of double lines in the fifth row are shown to coincide with the observed spectrum lines in the centre row. Drawing by CPS (Trans. RSE vol. 32).

wavenumbers. However, his work on other molecules went unrecognised: as Kayser said, 'who would have the patience to convert 1625 lines from reciprocal British inches?'

In April 1882 P G Tait discussed with Piazzi Smyth an experiment which he himself had tried unsuccessfully but which he thought might possibly be solved by Piazzi with his superior spectroscopic equipment. Its purpose was to look for possible shifts in the positions of spectrum lines when the light source producing them was viewed from different directions. The question was whether the ether which reputedly carried the light waves was stationary or in motion relative to the Earth. As Tait wrote to Thomson:[2] 'I have set Piazzi on to try his magnificient Gitter. If the ether be in motion relative to the Earth, the absolute deviations of lines in the diffraction spectrum should be different at different azimuths: unless of course the relative motions of Earth and ether be *vertical*.' In the experiment light was to be sent simultaneously into the spectroscope from two opposite directions by means of two sources and two reflectors. The apparatus was to be rotatable and the effect of a change in azimuth on the relative displacement of the spectral lines—if any—was to be looked for. Clerk Maxwell had tried the same experiment in 1868 using prisms, with a

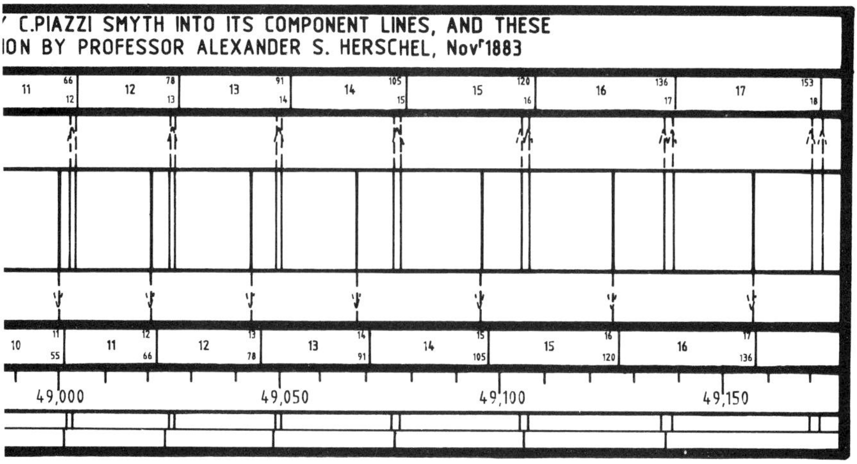

Alexander S Herschel (Science Museum).

null result.[3] Tait had also tried the experiment with his prismatic spectroscope, but its dispersion was not sufficient to give more than a 'suspicion' now and then of the effect. Piazzi Smyth heartily entered into the project and outlined his plan to Tait who responded 'Bravo! If after the innumerable screws are put in you get a result you will be entitled to rank as joint discoverer.' The entire solar spectroscope was made to revolve on a pin in the middle of the laboratory floor and preliminary astronomical observations were made to determine the exact orientation of the laboratory. Piazzi Smyth took great care to produce the narrowest possible spectrum lines: the sodium D lines from a Bunsen burner were very broad and hazy, but with hydrochloric acid vacuum tubes they showed up (presumably through contamination) as faint but beautifully fine. The observations, made with the spectroscope placed parallel and at right angles to the direction of the Earth's orbital motion, gave a null result within the errors of observation which he estimated as a shift in absolute positions of the lines with varying azimuth of 1/200 of the separation of the two sodium lines which are six angstroms apart. This placed an upper limit on the effect. These interesting experiments were unfortunately never published and Tait's biographer was not aware that they had ever been made. The crucial experiment on the luminiferous ether was, of course, the celebrated Michelson–Morley experiment of 1887.

It is worth pointing out that the manner in which Piazzi Smyth was able to define the degree of uncertainty in his result contrasts with his totally different approach to the interpretation of his pyramid metrology where he accepted the predicted theoretical answer as infallible and looked upon the observations as simple approximations to it, rather than the other way around. Those critics who looked only at his pyramid work may have concluded that all his scientific research suffered from the same lack of judgement.

The year 1880 added a new chapter to the saga of the metrology of the Great Pyramid with Flinders Petrie going out to Egypt to continue Piazzi Smyth's work. His Egyptian adventures followed closely those of the Smyths 15 years earlier. He was met in Cairo by Alexander Grant, the Scots doctor who had befriended the Smyths; he got to know the archaeologist Marietta and acquired as his personal helper at the pyramid the same Alee who had served the Smyths and before them Howard Vyse, and who in the meantime had also served Waynman Dixon, the engineer who had been introduced to Egyptology by Piazzi Smyth. Like the Smyths, Petrie lived in a rock tomb on the site.

It did not take Petrie long to discover that the famous Earth-commensurate and divine cubit of 25.025 inches had no real existence. He wrote to Piazzi Smyth and arranged to have his sheets of measurements in the pyramid's antechamber sent to him. These new findings were obviously not to Piazzi Smyth's liking, but he commented in his diary: 'Letter from Flinders Petrie. If I have been backward and supine in the cause, may be succeed!' Consumed with curiosity to know how Petrie was progressing he later wrote to ask Dr Grant, who was puzzled that Piazzi Smyth had not asked Petrie directly. For his part Petrie always acknowledged his debt to his former mentor giving him full credit for his assiduous labours and quoting his pyramid measurements throughout the latter's life and well after it. Petrie's scholarly article on the pyramids in the eleventh edition of the *Encyclopaedia Britannica* and reprinted until the fourteenth edition of 1937, refers to Piazzi Smyth's metrology—but with no mention of his mystical theories.

Petrie's final results, published in his book *The Pyramids and Temples of Gizeh* in 1883, showed that the dimensions of the pyramid were seven times forty cubits for the height and eleven times the same quantity for the side. The simple ratio 22/7 together with the length of the normal cubit replaced the perfect π and the imagined sacred unit. All theorising about astronomical information such as the number of days in the year or the period of precession alleged to be contained in the pyramid were clearly shown to be erroneous. Yet, as Petrie says in his autobiography published 50 years later, 'the fantastic theories are still poured out and the theorists still assert that the facts correspond to their requirements. It is useless to state the real truth of the matter as it has no effect on those who are subject to this type of hallucination.' It was unfortunately only too true that his theory was dearer than the fact to Piazzi Smyth, yet there had been a time when Flinders Petrie and his father had wholeheartedly concurred with most of Piazzi Smyth's ideas. Piazzi refused to countenance the Petrie findings, and attacked them in his last edition of *Our Inheritance*, published after he retired. He could not understand how Petrie could renege on the inspiration theory while pronouncing that the pyramid incorporated in its dimensions the ratio of the circumference to the radius of a circle and while marvelling at the precision of its workmanship, 'equal to opticians' work of the present day, but on a scale of acres instead of feet or yards of material'. Considering that Piazzi Smyth's theories were founded on the basic premise that the pyramid builders knew the ratio of the circumference to the radius of a circle, it is perhaps understandable that from his point of view Flinders Petrie's apparent about-turn was illogical. The difference between the interpretations of Petrie and Piazzi Smyth was that whereas Piazzi Smyth was convinced that only supernatural guidance could account for the inexplicable skills of the pyramid builders, Petrie simply recorded what he found with an open mind. The unavoidable break with the Petries must have caused much sorrow to Piazzi Smyth. William Petrie had been the friend of his youth and Flinders the wonder child on whom they had pinned their hopes.

Piazzi Smyth found another thorn in his side at this stage. R A Proctor, well known as a populariser of astronomy, produced an elaborate theory of his own about the pyramid as a counter attraction.[4] He saw the pyramid's sloping passage as a meridian-orientated observatory of some kind, and incorporated his ideas into his popular lectures, where, as Piazzi Smyth put it, 'he preached his malversations of my writings'.

However, Piazzi Smyth still had many followers, some from the ranks of American anti-metre campaigners who set up an 'International Institute for Preserving and Perfecting Weights and Measures', based in Ohio and run by engineers on the new American Railways. With these men Piazzi Smyth was now back to where he started, defending the Imperial system against the French. It was under their influence that in his later years he insisted on using only British inches in his spectroscopic work where previously he had given his results in both inches and angstoms. When Flinders Petrie's book on the pyramid appeared the American organisation presented Piazzi Smyth with a formal document, a 'resolution' declaring Petrie wrong and Piazzi Smyth right! In his final report on the work of the observatory Piazzi Smyth mentioned his American supporters as evidence of his achievement in metrology. It may be worth mentioning that amongst Piazzi Smyth's other American sympathisers was James Abram Garfield, twentieth President of the United States.[5]

Chapter 14

The Last Expedition

One day in 1882 Piazzi Smyth read an account in *Nature* of the improved diffraction gratings produced by H A Rowland at the Johns Hopkins University, Baltimore, in the United States and immediately set his heart on one. These gratings were the result of Rowland's construction of the almost perfect screw with which large concave as well as plane surfaces could be ruled with unprecedented accuracy. Piazzi Smyth asked Rowland if he could have from him the largest available concave grating but Rowland advised him that for his purposes a plane grating of ruled surface five inches wide and three inches high with 14 400 lines to the inch would be most suitable. Such a grating, costing $100 or £20, was delivered to Piazzi Smyth a year later and installed in his laboratory.

Piazzi Smyth used this high-quality grating in 1884 to obtain a remarkable set of solar spectra covering the entire visible range. Though he had plans to introduce photography into his spectroscopy, Piazzi Smyth decided against it for the time being. The technical details were not easy to fulfil for someone working all alone as he was, and in any case he was extremely experienced and fast at visual work; on one occasion he recorded the positions of 1865 lines in the space of three hours, a rate of ten per minute. Furthermore photography had now become fashionable with solar spectroscopists who thus concentrated on the blue and violet wavelength regions to the neglect of the others which were not yet readily amenable to the labour-saving techniques of photography.

What prompted Piazzi Smyth to undertake one last solar expedition was the occurrence in 1884 of some extraordinary sunsets and the reports from all over the world of green and blue suns and gorgeous colours in the sky. These phenomena were put down to the effect of dust from the great volcanic eruption of Krakatoa in the Indian Ocean in September 1883 in which half the mountain's bulk was

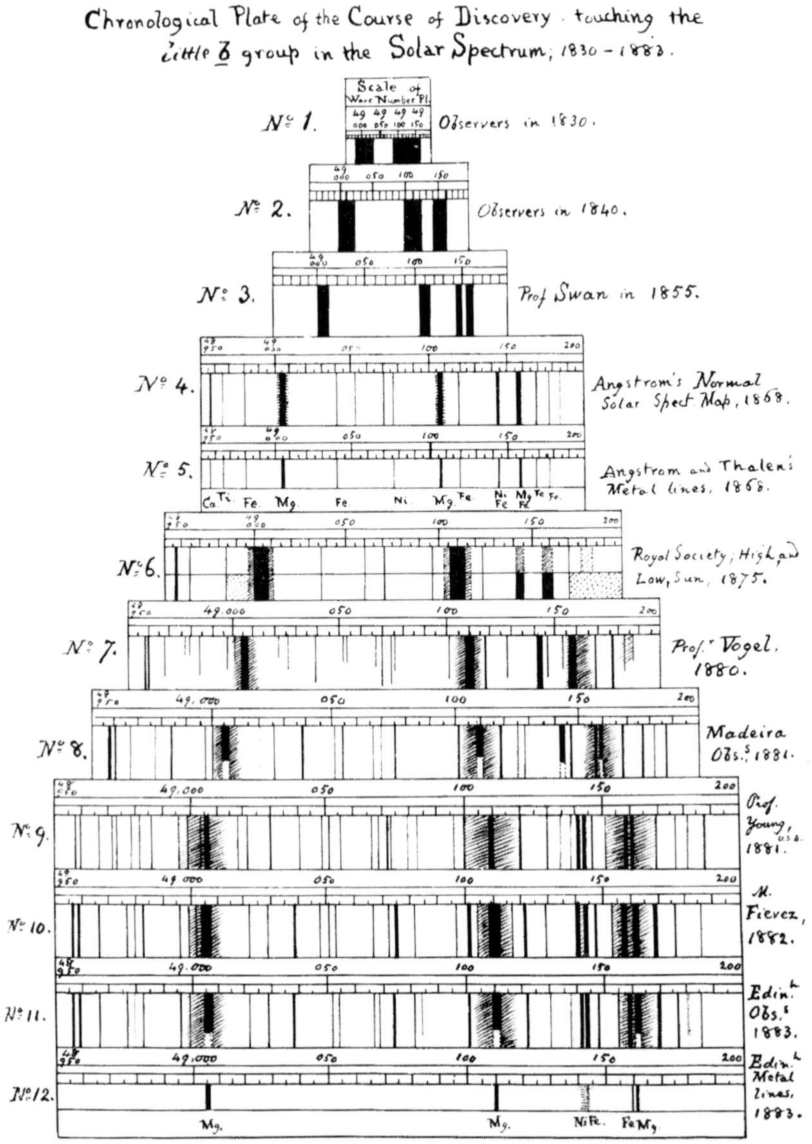

The improvement in spectroscopy from 1830 onwards illustrated by various spectra of a small (20 angstrom) sample region of the solar spectrum (CPS, Trans. RSE vol. 32).

estimated to have been blown away in the form of fine dust to a height of 40 miles in the atmosphere. Samples of the dust had been found in many quarters of the globe. Piazzi Smyth, recalling the death of Pliny from noxious gases in the famous eruption of Mount Vesuvius in

79AD, suggested that the Krakatoa dust clouds might also be accompanied by gases from the Earth's interior which might perhaps be different from the known gases in the normal atmosphere. It struck him that these gases might well be detectable spectroscopically in the spectrum of the Sun when compared with its spectrum in normal circumstances.

With his wife now an invalid there was no question of an expedition to a foreign shore, so he did the best he could and travelled to the sunny south of England, where he was able to rent a house a few miles outside Winchester, 'once the ancient metropolis of England under her Saxon Kings'. Ever optimistic, Piazzi declared the clean air of the countryside suitable 'for both solar originated lines and those produced by the higher telluric atmosphere whether natural or adventitious'. It had taken him several months to construct, single-handed, a large transportable metal mounting for his spectroscope for which the opticians Cooke of York supplied four-inch collimator and telescope lenses with focal lengths of 70 inches. With the new Rowland grating the spectroscope gave a dispersion of four angstroms per millimetre in the second order. This was a very large instrument, and it was a major undertaking for one person to pack, transport and reassemble it on his own. He had also added an oscillating lens of the kind which Cornu had mounted in front of his spectroscope slit to send the light from the edges of the Sun alternately into the instrument in order to distinguish the telluric from the solar lines, but he was not satisfied with it and did not use it. The apparatus was set up in a south-facing room in the house in Winchester. Since Jessica was by then incapable of performing her usual task of operating the heliostat Piazzi relied for assistance on a neighbour in Winchester, Colonel E Knight, an enthusiastic amateur astronomer. Though Piazzi had to nurse his wife and walk daily into the town to fetch her medicines, their combined efforts still led to the recording of the entire visual solar spectrum from 4040 to 7700 angstroms three times in only two months of that summer.

No individual observer since Ångström had recorded the whole visible range of the solar spectrum. The particular value of Piazzi Smyth's observations was that he had been able to record the wavelengths of almost 7000 lines with unsurpassed accuracy and to display them in maps which indicated symbolically the width and depth of each line. The precision of their positions was guaranteed by the consistency of the three independent sets of spectra among each other; comparison with the modern revised Rowland tables confirms the accuracy of Piazzi Smyth's wavelengths and shows that the resolution of the Winchester spectra was better than a tenth of an angstrom and that lines with equivalent widths as small as six or seven milliangstroms were detected on the best spectra. For comparison Piazzi Smyth reproduced Ångström's standard spectrum and the spectra from two famous sources, the blue and violet from Vogel's photographic spectrum of 1880 and the remainder from Fievez's visually observed spectrum of 1882. There is no doubt that Piazzi Smyth's spectra were superior to both.

The stated purpose of the observations—the detection of Krakatoa gas features—was not achieved, however. To complete the experiment Piazzi Smyth would have needed to repeat his observations under identical conditions after a lapse of time, but this did not prove feasible. All he could say was that there were no conspicuous new features in the solar spectrum when compared with previous versions, and he added the date, 1884, to his own observations for possible future comparisons. An attempt to detect possible changes in the continuous spectrum was also inconclusive.

In recording the positions of spectrum lines Piazzi Smyth had always favoured the use of wavenumbers or inverse wavelengths in preference to wavelengths. When it came to publishing the Winchester results in the *Transactions of the Royal Society of Edinburgh* Piazzi Smyth foolishly insisted on using a scale of inverse British inches on his charts; only at the last moment was he persuaded by James Gordon, the Librarian of the Royal Society of Edinburgh and a great admirer of his, to add a scale of wavelength in angstrom units. This Piazzi Smyth conceded with some reluctance, dubbing the units as 'wavelengths in modern French terms, adopted by Ångström in the latter years of his life, vice the "inches" of his renowned and heroic Scandinavian forefathers'. Thomas Heath, who had prepared the charts and had done all the preliminary computations, added the angstrom scale by hand.

The 60 beautiful charts were—quite unnecessarily and at Piazzi Smyth's own expense—reproduced in colour, with the spectrum divided into 12 colour bands. This was the result of his current enthusiasm for colour coding and was certainly a retrograde step: the whole purpose of using a wavelength scale, as Piazzi Smyth himself was one of the first to emphasise and had so eloquently demonstrated on his Palermo trip, was that the eye's impression was replaced by an exact numerical datum. 'But no matter what strange, artificial and human devised scale any one may have been using he must also, if an

(Opposite) One of the 60 sections of the Winchester solar spectrum, recorded three times and shown with other observers' spectra for comparison. This part covers 40 angstroms in the blue part of the spectrum from a total of 3700 angstroms recorded (Trans. RSE vol. 32).

46

THE VISUAL SOLAR SPECTRUM IN 1884.

illustrated by three separate Solar Spectra independently observed at Winchester in June and July, with a Rowland's Grating Spectroscope, by C. Piazzi Smyth, F.R.S.E.

two references to older Solar Spectra by earlier observers, being prefixed; and all of them reduced to a uniform Scale of Number of Waves of Light per British inch.

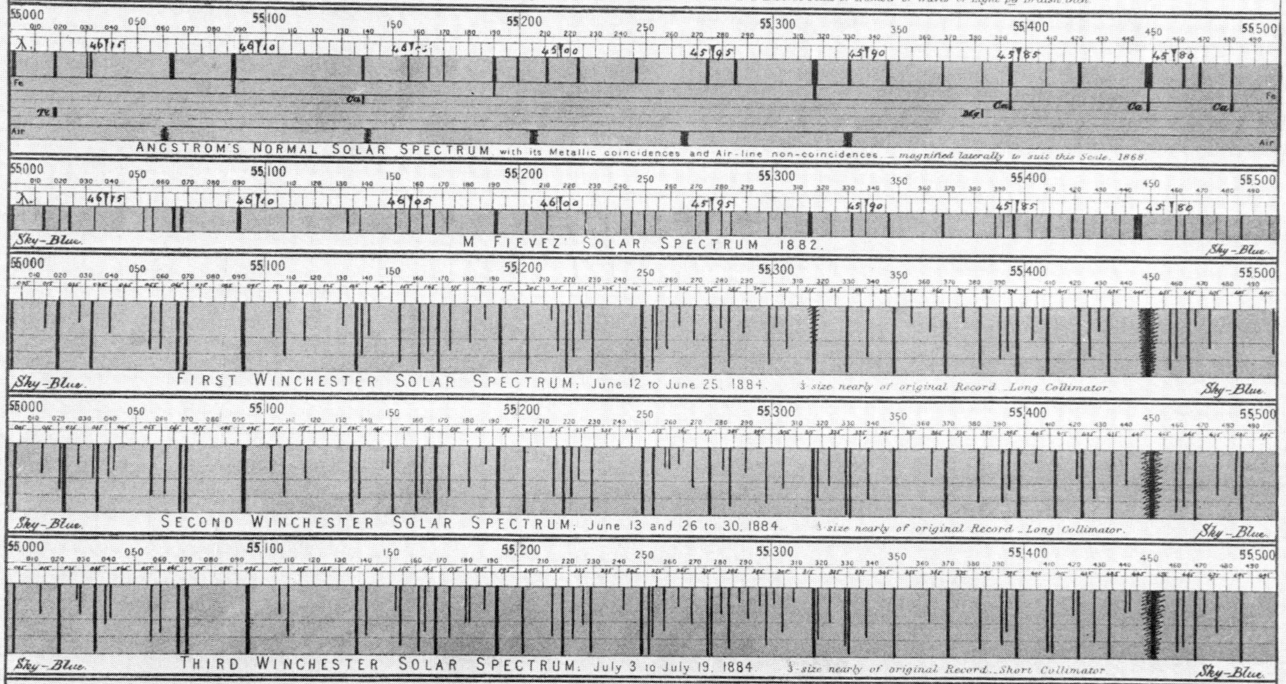

GENERAL RULES FOR THE METHOD OF REPRESENTATION ADOPTED IN THE ABOVE SPECTRA.

① A Vertical line always stands for a veritable, seen and measured Spectroscopic line or image of the Slit, and for nothing else under the Sun.

② Lines in any other direction than vertical, whether horizontal, or slanting at any angle and from either side, are to be interpreted as nebulous shade only, in vertical bars or bands at the place.

③ Greater or less height or depth, either of lines, or of shaded bands, _ is intended to typify greater or less intensity of darkness and visibility of such lines or bands.

④ Cones of shade arranged on a vertical central axis, indicate nebulous bands shaded off towards either side very gradually and delicately.

observer, have had Nature's inimitably beautiful and effective general indexing of spectrum place by colour before his eyes again and again', he declared. Some traditional spectroscopists, however, appreciated the coloured maps; Copeland, an old hand at visual spectroscopy of comets and nebulae, wrote from Dunecht: 'I for one am very obliged to you for the colour, for it makes me feel quite at home with the spectra.'

Piazzi Smyth read his paper on solar spectroscopy by invitation to the Royal Society of Edinburgh in the following year, 1885, when he also exhibited some of David Gill's star photographs. This paper and the coloured charts were bound and distributed from the Royal Observatory to 177 addresses, all but 21 within Britain and the 21 to 'British possessions'; in other words, not a single copy of this very important map of the solar spectrum reached the continent of Europe or the United States, with the exception of those few institutions which subscribed to the Society's *Transactions*. The reason for this limited circulation was the expense of printing. Had Piazzi Smyth been content to reproduce his spectra in a simpler form, and had he simply identified the known spectrum lines and listed the new ones, he would have done a great service to the solar spectroscopists of the time. As it was, the excellent Winchester spectra did not receive the attention which they certainly deserved; they are not mentioned in any historical accounts, not even in Kayser's exhaustive *Handbuch*.

The much discussed meteorological effects which followed the Krakatoa eruption engendered a great interest in watching the sky for unusual phenomena. Among these were noctilucent or night-time bright clouds which occur in the upper atmosphere at a height of about 80 kilometres, high enough for the clouds to be illuminated by the Sun and to be seen when the Sun is well below the horizon. The geometry of the situation means that the clouds become visible in midsummer in a zone of mid to high latitudes, which in the northern hemisphere includes Great Britain. Modern observations show that the clouds are composed of minute water-ice crystals which form under special conditions in the very cold high stratum of the atmosphere.

The first to record noctilucent clouds, in the summer of 1884, was a nature lover in the south of England and among those who made sightings in the summers that followed was Piazzi Smyth, who described them and their spectra in a letter to *Nature* in August 1886. In fact Piazzi Smyth had observed similar bright clouds in 1882 which were so conspicuous and unusual that he made them the subject of a major paper read before the Royal Society of Edinburgh. He described how, when on an evening in April of that year he prepared to make his customary nightly inspection of the sky from his north-facing auroral laboratory, he saw an 'unbelievable' sight—white clouds in a dark but starry sky. The Moon had not yet risen and there was no Aurora. Amazed, he went out of doors and found that there were bright clouds in the south part of the sky as well, 'luminous with a white, moon-like radiance' such that 'tree tops projected on them showed the details of twigs with almost daylight minuteness'. He examined the strange 'cottony' clouds with a pocket spectroscope and saw a faint continuous spectrum in the green. Deciding that the source of illumination could only be the gaslights in the city of Edinburgh he interpreted the bright appearance of the clouds as being caused by small water drops illuminated from below. When he searched the Scottish meteorological records for clues, he found that a wave of unusual dryness could be traced right across Scotland which passed over Edinburgh at the time when the luminous clouds were manifest, and he also inferred from the data that the clouds were at a height of 2000 feet or more. He suggested that under these abnormal conditions the atmosphere just below the clouds could be compared to the atmosphere just above them as observed in Tenerife. It is impossible to tell whether this explanation was the true one but the episode is worth recording as an illustration of Piazzi Smyth's gift of observation and of his thoroughness in seeking answers. The phenomenon resembled noctilucent clouds, but according to a modern expert, Michael Gadsden of Aberdeen University, the position of the clouds in the sky and the time of year rule out this possibility.[1] Nevertheless Piazzi Smyth, with his sighting in 1886, remains one of the very first observers of genuine noctilucent clouds.

In 1884 the University of Edinburgh celebrated the tercentenary of its foundation with a week-long celebration in April. Each professor in the university had been asked to recommend two candidates for honorary degrees, and foreign universities and academies had been invited to send representatives. Piazzi Smyth records that A S Herschel represented him as the delegate of the Royal University of Palermo. The honorary LL D degrees recommended by Piazzi Smyth went to two outstanding astronomers, one representing theoretical and the other practical astronomy. They were Simon Newcomb, the doyen of American astronomers and Director of the American

Nautical Almanac Office, and David Gill of the Cape. There was insufficient time to arrange for Newcomb to appear in person, but Gill was present, on his first visit home in the five years since his appointment to the Cape. It undoubtedly gladdened Piazzi Smyth to see his friend again, by now an internationally acclaimed astronomer. David Gill certainly enjoyed the occasion when, as he wrote to Newcomb, he had 'worn a red gown and a velvet cap, and so been transformed into a Doctor of Laws—what kind of laws I am learned in I have yet to ascertain'.

It was for this occasion of the tercentenary that the Principal of the University, Sir Alexander Grant, wrote his *Story of the University*, a history of the university and of the individual Chairs. The Chair of Astronomy, and particularly Piazzi Smyth's tenure, were treated in this history in a far from sympathetic way. Having summed up the findings of the Government Commission, quoting Piazzi Smyth's objections to university teaching duties, Grant's account continued: 'No doubt the Astronomer Royal is right in keeping the profanum vulgus out of his "Uranienburg", and it would perhaps be unreasonable to expect him to give courses of lectures in the University. But, after all, the Observatory, as well as the Professor of Practical Astronomy, belongs to the University', and ended: 'It must be said that the Chair of Practical Astronomy, which has now been in existence nearly a hundred years, has contributed next to nothing to the educational resources of the University.' The tone of Grant's account was unwarrantably sarcastic and very hurtful to someone who was still a member of his own Senatus. Grant being a member of the Observatory Board of Visitors ought to have known how hard Piazzi Smyth worked and how difficult he found it to keep the establishment running on inadequate government funds, and he ought also to have remembered that in all the years of Piazzi Smyth's tenure the university had made just one single contribution of £10 towards students' equipment!

Piazzi Smyth was to teach one last student in the university. In the autumn of 1885 he received a request for instruction in astronomy from a young American, Joseph Swaine, Associate Professor of Mathematics at Indiana University, who arrived with a letter of recommendation from David Kirkwood asking him 'to do immense things' for him. Piazzi Smyth felt that there was very little he could teach him, but Swaine was determined to be accepted. In fact he spent a full year at Edinburgh University studying under Tait and other professors as well as being taught practical astronomy by Piazzi Smyth and Heath. Swaine returned to the United States to a very successful career, becoming successively President of Indiana University and of Swarthmore College.

The programme in positional astronomy which Piazzi Smyth began in 1849 continued unabated for over 20 years during which time more than 25 000 star positions were measured with the transit instrument and the mural circle. Piazzi Smyth, with his main interest in physical astronomy and his taste for travelling, was not always available to take his share in the observations, and the main burden had fallen on the devoted Wallace who completed most of the reductions before he retired in 1880. The planned programme was carried through for 20 of the originally contemplated 30 years and the last of the observations covering the years up to 1870 were published, after delays and difficulties with the Treasury about printing costs, in the *Edinburgh Astronomical Observations*.

Piazzi Smyth decided that there was now a sufficient body of accumulated material to justify combining it in a large catalogue which might reveal the stellar movements which the programme was intended to discover. In this catalogue one whole page was allocated to each star listing the various observations made at different times, reduced to the positions of the equinoxes of 1830, 1850 and 1870. In addition to the Edinburgh observations, those made at other 'reputable' observatories—nine in number—were added, reduced to the same equinoxes. Proper motions according to other published catalogues were also listed in order to provide as accurate values as possible and reveal possible anomalous motions. The amount of labour involved in performing these additional reductions not only for the Edinburgh observations but for those of other observatories was immense and must have sorely tried the endurance of Piazzi Smyth's assistants. The catalogue, the last of the series of the *Edinburgh Astronomical Observations*, was published in 1886. It contained all known reliable positions of a total of 3890 selected stars, between magnitudes two and ten. With its completion, Piazzi Smyth had carried out his pledge to produce a comprehensive systematic body of star positions over a significant period of time.

Unfortunately the Edinburgh star catalogue did not fill the historic role that Piazzi Smyth had hoped. He himself, his heart more deeply in his spectroscopy than in the stars, made no attempt to analyse the data contained in it. By the 1880s the problem which the catalogue was to solve was being approached more directly, by comparing entire catalogues of earlier epochs with the most recent observations. Bradley's Greenwich catalogue of 1755 was re-reduced by Auwers of Berlin and compared by Ludwig Struve of Pulkova with the 1855 Pulkova catalogue in 1887, almost simultaneously with the appearance of Piazzi Smyth's final volume. 'No doubt the chances of success for such a gigantic undertaking were then extremely small', wrote one of the Edinburgh astronomers twenty years later,[2] 'nor was the greatness of the task in proportion to the limited resources available to its originator, so that the practical result attained by the publication of those two volumes was perhaps small; yet we may look upon Smyth's compilation as the first, however crude, attempt at a "history of the fixed stars".' There was, however, continuing demand among astronomers for accurate early star positions, and Piazzi Smyth's successor, Ralph Copeland, on the advice of the leading positional astronomers of the time, decided that it would be worth re-reducing the original Henderson–Wallace catalogue of observations made in the 1840s. The new revised version of the Henderson catalogue which had originally been brought out by Piazzi Smyth in 1850, was published in 1906.

Another laborious task which Piazzi Smyth, with his obsession for committing to print for the benefit of posterity the details of every scientific observation he ever made, undertook during his last years in Edinburgh was to assemble the meteorological data for Scotland which the observatory had collected since 1858. The observatory was without a second assistant, Treasury having refused to replace the previous one, and Piazzi Smyth actually paid a young assistant out of his own pocket to help with these calculations. Various tabulations of the meteorological data, covering 32 years, and also the remaining unpublished readings of the rock thermometers, were published in two massive papers, occupying 128 pages of the *Transactions of the Royal Society of Edinburgh*, and seen through the press by Thomas Heath in 1890.

One matter which haunted Piazzi Smyth in his last years in office was the problem of the failed equatorial, and before laying down his burden he was determined to have the last word on this unfortunate venture. He delivered a long address to the Royal Society of Edinburgh at their December meeting of 1887 on 'The Edinburgh equatorial in 1887' in which he explained the history of the project, the reasons for its apparent failure and his remedies for turning it, even at this late stage, into a useful instrument. His recommendations

(Opposite) A satirical drawing by Piazzi Smyth showing the state of the equatorial telescope in its cramped dome in 1887, blamed on the Office of Works, and 'some of the many successive positions of the observer in order to accomplish an observation'.

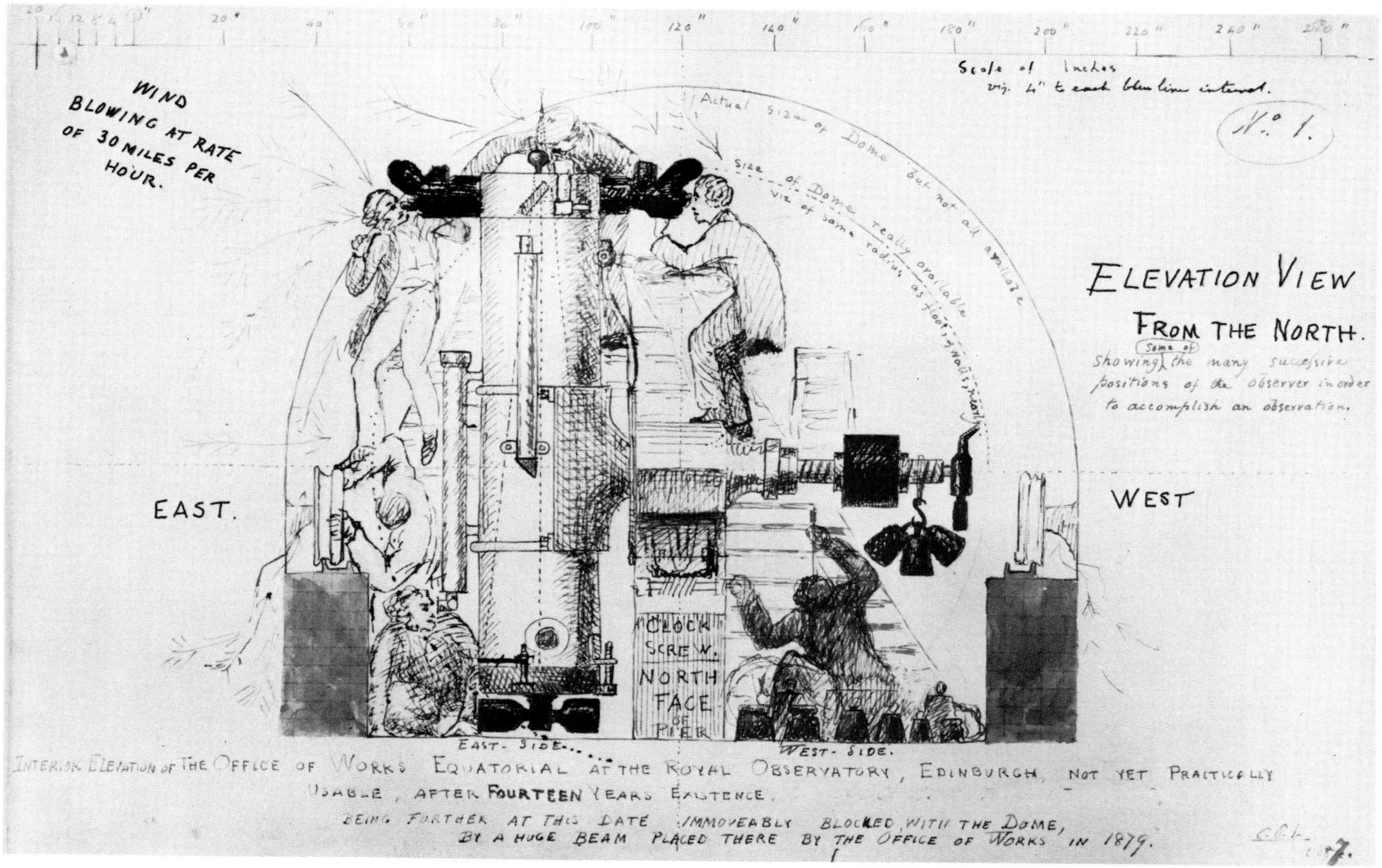
20" 40" 60" 80" 100" 120" 140" 160" 180" 200" 220" 240" 260"
Scale of Inches
viz. 4" to each blue line interval.
No. 1.
WIND BLOWING AT RATE OF 30 MILES PER HOUR.
Actual size of Dome but not all available
Size of Dome really available viz of same radius as floor, & walls, nearly
ELEVATION VIEW
FROM THE NORTH.
Showing some of the many successive positions of the observer in order to accomplish an observation.
EAST.
WEST
CLOCK SCREW.
NORTH FACE OF PIER
EAST-SIDE.
WEST-SIDE.
INTERIOR ELEVATION OF THE OFFICE OF WORKS EQUATORIAL AT THE ROYAL OBSERVATORY, EDINBURGH, NOT YET PRACTICALLY USABLE, AFTER FOURTEEN YEARS EXISTENCE.
BEING FURTHER AT THIS DATE IMMOVEABLY BLOCKED, WITH THE DOME, BY A HUGE BEAM PLACED THERE BY THE OFFICE OF WORKS IN 1879.

for improvement were, however, tantamount to a complete rebuilding of the telescope which may have made his audience wonder why it had not been designed differently in the first place. The paper included an airing of his recurrent grumbles about the low level of support of the Royal Observatory at Edinburgh and the low salaries of his staff. As a farewell offering to the Royal Society of Edinburgh the paper did not do much credit to Piazzi Smyth who by that time had become decidedly eccentric. However, the Society's Vice-President, Lord Maclaren—like his late father Duncan Maclaren a stalwart supporter of Piazzi Smyth—who introduced the paper made a strong plea for the greater support of the observatory. Countering the argument that the country did not need two national observatories, he stressed that 'there is scope for division of labour in astronomy as in other things. It is really not doubtful that, with proper equipment, our Observatory could be employed on work such as can only be done at a national observatory, under official rules and discipline, which should not in the least compete with the work done at the great National Observatory at Greenwich'. A very sympathetic view of Piazzi Smyth's complaints was also taken by the editor of *Nature* who declared that Piazzi Smyth could not be blamed for deciding to retire under the circumstances.

The equatorial telescope was never used on Calton Hill, and when after Piazzi Smyth's retirement the observatory was moved to its new site on Blackford Hill pride of place in the large new dome was given by the new Astronomer Royal, Ralph Copeland, to the 15-inch refractor of Dunecht, the splendid telescope which Grubb had made for David Gill at the very same time as they constructed the ill-fated 24-inch reflector for Piazzi Smyth. The equatorial was relegated to a smaller dome and though Copeland gave a flattering account of it to the press in his report on the new observatory, he clearly realised its faults from his earlier examination. The telescope was never used for any serious work, and when eventually its faults were critically examined by R A Sampson in 1910 the instrument was finally discarded.[3] The real reason for the trouble was that the state of the telescope-makers' skill was not sufficiently advanced in 1871 when Piazzi Smyth planned his ambitious portable equatorial. He was not to know then that the much lauded four-foot reflector made by Grubb for the Melbourne Observatory would turn out to be nothing less than 'a disaster'. In a review of 'monster telescopes' D E Osterbrock[4] shows clearly that the day of the large reflectors did not come until the turn of the century.

In Piazzi's last years in office Jessica was constantly ill in spite of visits to brine baths and the attentions of doctors. The anxious Piazzi took

The 24-inch equatorial telescope when re-erected at the Royal Observatory on Blackford Hill at the end of the century (ROE).

hygrometer readings wherever he went, hoping to find a suitable dry location for his wife's permanent residence. He was impressed by the Malvern Hills which seemed dry and healthy and he contributed, in fact, a short paper on 'Malvern, queen of inland health resorts' to the British Association. While looking for a suitable house in the area the couple visited the small Yorkshire town of Ripon—'the finest day we have had since we came here', wrote Jessica in her meteorological journal, one of the very rare occasions when she recorded anything beyond the strictly scientific. Their one day's visit decided them on acquiring a house in Ripon which became their second home from December 1886.

Even in her greatest distress Jessica faithfully kept up her meteorological journal. She recorded twice-daily readings of barometer, wind velocity and direction, thermometer and humidity, and after Piazzi's discovery of the rainband her records also included the strength of this band on a scale of one to ten, cloudiness on the same scale, the number of hours of sunshine and the occurrence and intensity of rain or snow. For ten years, from 1876 to 1886, Jessica carried her little meteorological laboratory with her wherever she went, including the trips abroad and to health resorts. When a new sunspot cycle came around, she added drawings of the spots as well. When her state of health did not allow her to move beyond her conservatory Piazzi began to illustrate her journal with drawings of their rare plants, some of which had been brought home as cuttings from Madeira. He also took a series of photographs, which still survive, showing the growth of ferns indoors. Jessica's meteorological journal came to an end with the move to Ripon. It was an unbroken record spanning ten years but one which was unfortunately never put to the scientific use for which it was intended, namely, the study of the rainband as a possible tool in meteorological observations.

Jessica's share in her husbands's work deserves to be remembered. She was entirely happy in her role as a devoted and painstaking scientific assistant, but she was also a perfect secretary to her husband, copying his manuscripts, filing his letters and even looking after the observatory finances on occasion when Piazzi was too engrossed in his own work. Her small contribution to science, the meteorological journal, was a work of great devotion which brought her no rewards. The Smyths' friend in Tenerife compared her rightly to Caroline Herschel; Haliburton, Piazzi's friend, was moved to congratulate him on having 'a wife that can sympathise and work with you in your researches'; and when a French spectroscopist visited them in Edinburgh he referred to 'the most intelligent and agreeable assistant in the person of Madame Piazzi Smyth'.

Piazzi was naturally anxious to relinquish his observatory post at this time in order to stay with his invalid wife. However, when he intimated his intention to the Home Office he faced the problem of the lack of any provision for a retirement pension. He was also resolved to leave the affairs of the observatory in good order and not to burden his successor with unreduced and unpublished observations, as had been left to him when he arrived in Edinburgh 40 years earlier. The question of a pension brought up the old thorny issue of whether his joint post was primarily a university or a government one but after a good deal of wrangling the matter was at last satisfactorily settled and Piazzi Smyth was granted a generous pension of £300 a year.

Before finally leaving Piazzi Smyth wrote and printed a report on the state of the observatory which he addressed to the Marquis of Lothian, the Secretary for Scotland. In this he described what he considered the most important achievements during his time in office. He referred first and foremost to his great star catalogue which he was proud to have accomplished, the publication of which, he said, made the moment opportune to offer his resignation 'and leave new hands

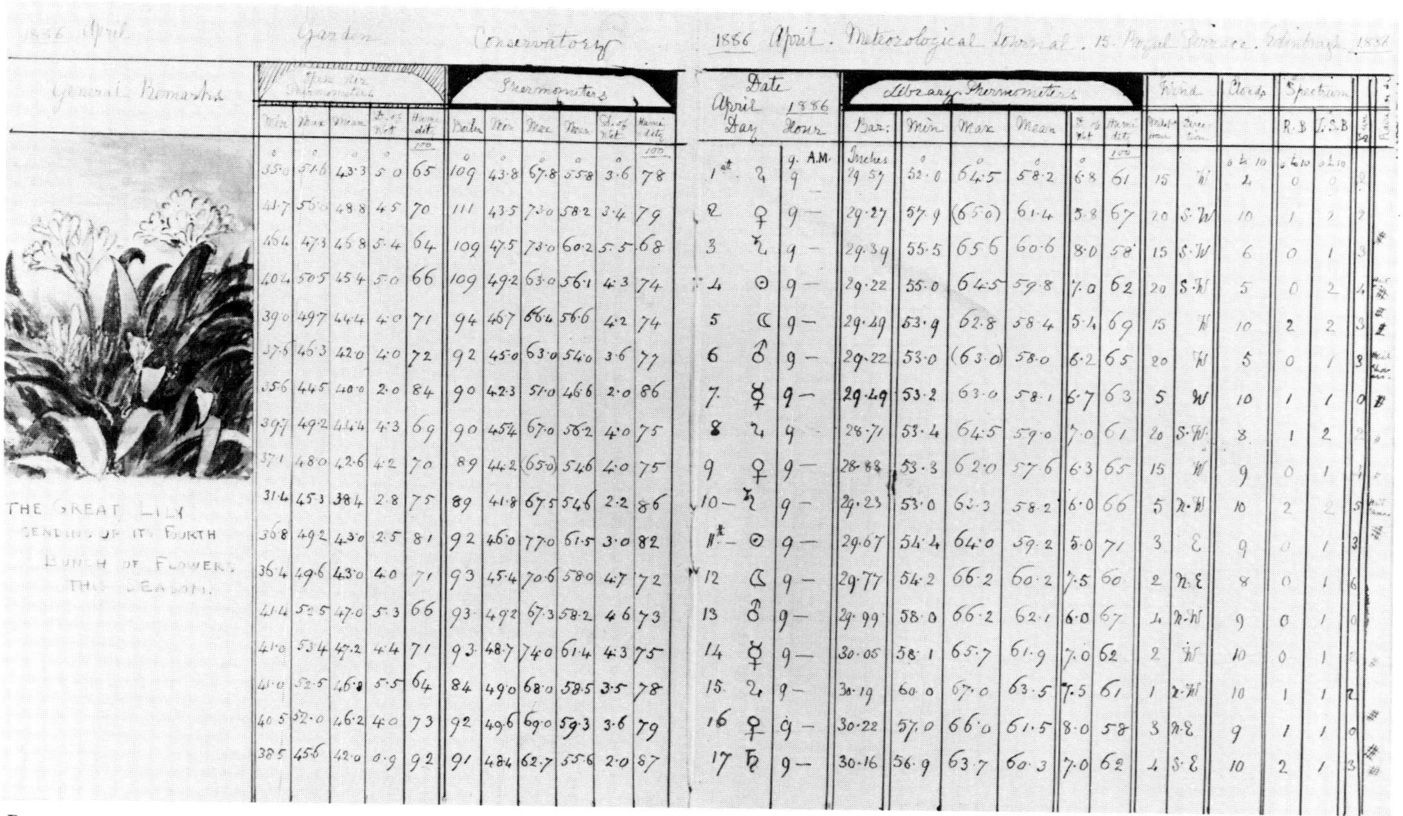

1886 April — Garden — Conservatory — 1886 April. Meteorological Journal. 15 Royal Terrace. Edinburgh 1886

General Remarks: THE GREAT LILY SENDING UP ITS FOURTH BUNCH OF FLOWERS THIS SEASON.

Garden (Open Air Thermometers) Min	Max	Mean	Diff. of Wet	Humidity	Conservatory (Thermometers) Bulb	Min	Max	Mean	Diff. of Wet	Humidity	Date April Day		1886 Hour	Library (Thermometers) Bar.	Min	Max	Mean	Diff. of Wet	Humidity	Wind Velocity	Direction	Clouds	Spectrum R.B	D.S.B
°	°	°	°	100	°	°	°	°	°	100				Inches	°	°	°	°	100			0 to 10	0 to 10	0 to 10
35.0	51.6	43.3	5.0	65	109	43.8	67.8	55.8	3.6	78	1st	♃	9 A.M.	29.57	52.0	64.5	58.2	6.8	61	15	W	2	0	0
41.7	55.0	48.8	4.5	70	111	43.5	73.0	58.2	3.4	79	2	♀	9	29.27	57.9	(65.0)	61.4	5.8	67	20	S.W	10	1	2
46.4	47.3	45.8	5.4	64	109	47.5	73.0	60.2	5.5	68	3	♄	9	29.39	55.5	65.6	60.6	8.0	58	15	S.W	6	0	1
40.4	50.5	45.4	5.0	66	109	49.2	63.0	56.1	4.3	74	4	☉	9	29.22	55.0	64.5	59.8	7.0	62	20	S.W	5	0	2
39.0	49.7	44.4	4.0	71	94	46.7	66.4	56.6	4.2	74	5	☾	9	29.49	53.9	62.8	58.4	5.4	69	15	W	10	2	2
37.5	46.3	42.0	4.0	72	92	45.0	63.0	54.0	3.6	77	6	♂	9	29.22	53.0	(63.0)	58.0	6.2	65	20	W	5	0	1
35.6	44.5	40.0	2.0	84	90	42.3	51.0	46.6	2.0	86	7	☿	9	29.49	53.2	63.0	58.1	6.7	63	5	W	10	1	1
39.7	49.2	44.4	4.3	69	90	45.4	67.0	56.2	4.0	75	8	♃	9	28.71	53.4	64.5	59.0	7.0	61	20	S.W	8	1	2
37.1	48.0	42.6	4.2	70	89	44.2	(65.0)	54.6	4.0	75	9	♀	9	28.88	53.3	62.0	57.6	6.3	65	15	W	9	0	1
31.4	45.3	38.4	2.8	75	89	41.8	67.5	54.6	2.2	86	10	♄	9	29.23	53.0	63.3	58.2	6.0	66	5	N.W	10	2	2
36.8	49.2	43.0	2.5	81	92	46.0	77.0	61.5	3.0	82	11th	☉	9	29.67	54.4	64.0	59.2	5.0	71	3	E	9	0	1
36.4	49.6	43.0	4.0	71	93	45.4	70.6	58.0	4.7	72	12	☾	9	29.77	54.2	66.2	60.2	7.5	60	2	N.E	8	0	1
41.4	52.5	47.0	5.3	66	93	49.2	67.3	58.2	4.6	73	13	♂	9	29.99	58.0	66.2	62.1	6.0	67	4	N.W	9	0	1
41.0	53.4	47.2	4.4	71	93	48.7	74.0	61.4	4.3	75	14	☿	9	30.05	58.1	65.7	61.9	7.0	62	2	W	10	0	1
41.0	52.5	46.8	5.5	64	84	49.0	68.0	58.5	3.5	78	15	♃	9	30.19	60.0	67.0	63.5	7.5	61	1	N.W	10	1	1
40.5	52.0	46.2	4.0	73	92	49.6	69.0	59.3	3.6	79	16	♀	9	30.22	57.0	66.0	61.5	8.0	58	3	N.E	9	1	1
38.5	45.6	42.0	0.9	92	91	48.4	62.7	55.6	2.0	87	17	♄	9	30.16	56.9	63.7	60.3	7.0	62	4	S.E	10	2	1

Page out of Jessica's meteorological journal, including rainband observations (in the last columns), which she maintained for ten years (ROE).

and head to direct a radically new Royal Observatory of Edinburgh powerfully equipped in the manner devised by the independent and very able Government Commission of 1876 and 1879; and to be then set afloat on a new career of usefulness'. Apart from the time service, Piazzi Smyth mentioned just two other fields of progress associated with his period of office, the Tenerife experiment and the scientific metrology of the Great Pyramid.

When on 30 July 1888 Piazzi Smyth's resignation was accepted the Marquis of Lothian wrote; 'It now devolves on me to convey to you the Secretary for Scotland's formal acceptance of your resignation of the duties of the post of Astronomer Royal for Scotland which you have filled with such distinction for the period of over 40 years since the appointment was conferred on you by the gracious favour of the Sovereign. The Secretary for Scotland is desirous that your connection with the public service should not terminate without placing on record his appreciation of the zeal and ability which have always characterised your discharge of laborious official duties: and he desires to take this opportunity of expressing his high sense of the value of the eminent services which in your official as well as in your private capacity you have rendered in the promotion of the cause of science to which you have devoted your life.'

The Senatus Academicus of the University on 27 July also paid tribute to him, to his surprise and pleasure: 'The Senatus resolve to record their sincere regret that the long and honourable connection of Professor Charles Piazzi Smyth, Astronomer Royal for Scotland, with the University has now been severed, and that the students of the University so rarely availed themselves of their excellent opportunities of studying under the guidance of so able and competent a teacher of Astronomy. They also desire to express their high sense of the great services rendered to Science by Professor Piazzi Smyth during the long period of forty three years during which he was Professor of Practical Astronomy, and they earnestly hope that health and strength may yet long be vouchsafed to him in his honorable and well merited retirement.'

Finally, the Registrar General joined in expressing appreciation for 'the great obligations under which you have laid the Government and the public, by undertaking and carrying on during so long a period, the important and laborious duty of preparing the Meteorological Statistics'.

On 15 August 1888, Piazzi Smyth inspected the observatory, paid the young assistant, and handed his First Assistant some written 'Parting words of advice', which began 'Watchfulness and general preparedness for anything unexpected and untoward!' At 3.45 he had a final word with Mr Heath and at 4.05 he handed over the keys of the observatory to him and signed the staff book for the last time.

Chapter 15

Sun and Clouds at Clova

After a lifetime of hectic scientific activity which saw troubles as well as successes it might he expected that on leaving Edinburgh Piazzi Smyth would wish to disappear into quiet retirement with his invalid wife, leaving the cares of the world behind. However, to give up was not in Piazzi's nature. He was as much an enthusiast on the day he retired as he had been when he first began his astronomical career more than 50 years earlier, and the days of retirement were in many ways very happy ones. He saw himself, like his hero Sir John Herschel in his later years, as 'freed from responsibility of either Government office or social and University public arrangements, and in that blessed liberty throwing off happiest corruscations of imaginative thought'. His great solar spectrograph had been moved from Edinburgh and installed in his house in Ripon; he noted in his workbook: 'We are now living with our belongings in the little house "Clova", Ripon, hoped to form our permanent home for the rest of our lives.'

Clova, called after the estate in Aberdeenshire where Jessica had spent her happy youth, was by no means a little house—it is now a nursing home—though it was less majestic than the one the Smyths had left in Playfair's Royal Terrace. It was spacious, with large windows overlooking the countryside and was close to the therapeutic baths on the outskirts of the old cathedral town. The shrubs and trees which still adorn the grounds are reputed to have been planted by Piazzi Smyth himself. Ripon had been chosen not only for its climate but also for its nearness to the city of York with its mainline train connections to north and south, just about half way between Edinburgh and London, and incidentally also the site of the workshop of the important optical firm of Thomas Cooke.

Piazzi Smyth had the next stage in his programme of solar spectroscopy already planned. Having investigated in Lisbon and

Piazzi Smyth in old age with a grand niece (ROE).

Madeira the main features of the solar spectrum in detail, and having recorded at Winchester the entire visible span of the spectrum from 7600 angstroms in the red end to 4100 angstroms in the violet, he was now determined to round off the work by studying the far-violet part of the solar spectrum including the great H and K lines of ionised calcium. This meant turning from visual observation to photography.

With the help of a skilful assistant—a joiner by trade and a 'philosopher and friend'—Piazzi Smyth adapted his visual instrument to photographic work by means of a wooden camera taking 3.3-inch plates. In his spectroscope he used a plane grating of high quality which H A Rowland had sent him in 1886 to replace the earlier one which Rowland did not consider good enough. With a new screw drive for rotating the grating supplied by Cooke's and a new object glass Piazzi now had a very good spectrograph, though its performance could naturally not equal that of the new 21-foot concave grating spectrograph which Rowland brought into operation in 1889. Needless to say he could not help being a little envious of the work of Rowland who later sent him a copy of his new map of the solar spectrum which had been recorded on 10 photographic plates each 19 inches long. However, he carried on with his own instrument, using it both for his work on the ultraviolet solar spectrum and for his spectrochemical analysis of gases in his end-on tubes.

In combining photography and spectroscopy Piazzi Smyth was bringing together two fields of work in which he was an expert. The coloured spectral maps of Winchester were forgotten as he devoted himself to improving the definition and clarity of his photographic spectra. He succeeded in producing his first test photographs of the visible violet solar spectrum, which he could compare with his visual records, in time for the 1890 meeting of the British Association in Leeds. The Association awarded him a grant to continue his research and set up a committee consisting of leading spectroscopists of the day—Schuster, Liveing and Dewar—with himself as secretary, to plan the next stage of the project. He also received grants of £100 from both the Royal Society of Edinburgh and the Royal Society of London with which he was now reconciled though he never rejoined its ranks. The Royal Society made further annual grants, the first time in his life that he had proper funding for research.

Following some further improvements in the spectroscope which included a large Barlow achromatic lens, ordered by the committee and constructed by Cooke's, Piazzi Smyth was able to produce a sizeable extension to his ultraviolet solar spectrum. His new photographic map extended from the point in the violet where the Winchester spectra had left off to the ultraviolet wavelength of 3565 angstroms. The quality of these spectra was outstanding; enlarged on paper in 70 sections with a carefully marked scale of wavelength (in British inches!), copies were presented to the Royal Society of Edinburgh in 1891, which, however, were not published on account of the prohibitive cost of reproduction. With Rowland's definitive spectrum about to become available, Piazzi Smyth's spectra had in fact become redundant. However, it gave him great personal satisfaction to have studied and recorded the entire range of the solar spectrum from end to end.

Another new grating which he had ordered, five inches square and ruled on Rowland's latest engine, arrived from J A Brashear of Allegheny in February 1892. It had a ruled surface of 3.85 inches and 20 000 lines to the inch. Piazzi Smyth found it considerably more brilliant than the two which he already possessed. He installed it in his spectrograph and obtained another series of far-violet spectra in the course of the following three months. These spectra, presented to the Royal Society of Edinburgh, also remained unpublished. The next intended stage in his programme was to photograph simultaneously the metallic line spectra and the solar spectrum in the ultraviolet for the purpose of identifying some of the numerous unknown solar lines. However, he did not manage to obtain a sufficiently powerful electrical supply to run the discharges in his ultraviolet vacuum tubes and was forced to give up this attempt. Piazzi Smyth's last recorded entry in his bulky spectrograph book is 25 April 1892 and the last preserved solar spectra were taken on 12 May. The meeting of the British Association in Edinburgh in the summer of the same year when he reported the completion of his solar project to his committee and exhibited his spectra was his last appearance at a scientific gathering.

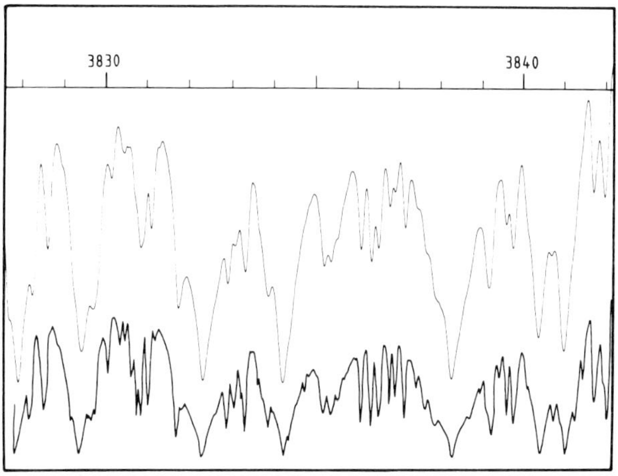

Microdensitometer tracing of a sample region of the Clova ultraviolet solar spectrum taken from the original photograph of May 1892, one of the last spectra taken by Piazzi Smyth. The same region from the Utrecht Atlas of the Solar Spectrum (1940) is shown (lower recording) to illustrate the high spectral resolution of Piazzi Smyths' photograph. The vertical scales are not comparable because the Utrecht Atlas is on a linear scale of light intensity. Wavelengths in angstroms (ROE).

Meantime, he returned to another project which had been dear to his heart for many years. This was the systematic photography of cloud formations intended as a scientific contribution to the general field of meteorology. When in 1876 he had been engaged in studying the rainband he had experimented with a 'cloud-taking camera' which produced negatives of clouds on small photographic plates. He was the very first to suggest photography as a serious tool for meteorological research, as the editor of *Photographic News* reminded his readers in 1888 when the idea of performing cloud photography was under discussion in Germany and the United States. When in his retirement Piazzi Smyth took up the matter again the words of John Herschel were once more in his mind, that 'meteorology is a science still in its infancy', words which had been spoken a long time ago, but were still largely true; in spite of the massive number of data collected world-wide by international cooperation the science of weather, which had always fascinated Piazzi Smyth, was still far from exact.

Piazzi Smyth constructed his cloud camera using a cherished Dallmeyer lens of 1.7 inches aperture and 14 inches focal length presented to him 30 years earlier by his good friend Joseph Sidebotham for the copying of his Egyptian pictures. With this camera he, assisted by his amanuensis Mr Close, obtained at Clova over a period of three years more than 500 photographs of cloud formations on three-inch square glass plates. Enlarged prints of 144 of these, illustrating every type of cloud formation, each accompanied by particulars of date, time and weather conditions, were assembled in three large albums which, with printed introductions, were presented to the Royal Society in 1897; another set of photographs and the original negatives went to the Royal Society of Edinburgh. These remarkable photographs which might have made a substantial contribution to the project of the International Meteorological Committee for a year of systematic cloud observation, including cloud photography, in 1896–7, were never put to scientific use.

The truth is that in the preparation of these beautiful photographs the artist had taken over from the scientist. Piazzi Smyth, now an old man, looked on his cloud forms as fleeting objects of beauty rather than as examples of natural phenomena. He called them 'Cloud forms that have been: to the Glory of God their Creator and the wonderment of learned men', and reflected on how each cloud, and each raindrop within each cloud, came about through the immutable laws of physics; yet no two clouds were exactly the same because the same identical circumstances could never recur. All his life he had loved looking at the sky and had recorded hundreds of skyscapes—sunsets, illuminated clouds, storms—with their endless variety of colours and combination of shapes. The photographs could capture only some of the many forms while most had gone unrecorded or unseen. 'Not one of them ever came back and showed its beauties again! Alas! Alas!, how many such opportunities have not I in the long life allowed me, and the varied course it granted, lazily and culpably lost!'

The cloud photographs were Piazzi Smyth's last work. It is fitting that they should have combined the scientist with the artist in him and that our last glimpse of him is free from controversy. In his regrets for opportunities lost he judges himself too harshly: the vast notebooks filled with scientific experiments, the sketchbooks and diaries, the numerous scientific papers, the hundreds of thousands of astronomical observations give a different picture, that of a man who never spared time or energy in the cause of science, who worked hard every day of his life at the exploration of the world around him, even if some of his activities led him into strange paths.

Two years after Piazzi Smyth retired to Clova the University of Edinburgh, under its new Principal Sir William Muir, awarded him the degree of Doctor of Laws Honoris Causa. The hand of his old friend P G Tait was undoubtedly behind the proposal which was initiated by the Faculty of Arts and adopted unanimously by the Senatus and Court. The degree was conferred *in absentia* in 1890, Piazzi being unable to leave his ill wife. The degree gave him all the more pleasure because it was his first and only university degree; thenceforth he was proud to add LL D to his other distinctions. It was a happy thought on the part of the University to make amends, after more than 40 often stormy years.

In 1896, the same year which saw the opening of Edinburgh's new Royal Observatory, Piazzi Smyth lost his beloved Jessica. It is moving to read in the preface to the first volume of the *Cloud Forms* the reference to himself as 'now in retirement at Clova with his dear wife Jessie', amended in the third volume to 'now in retirement alone through the deeply lamented death of his dear wife Jessie'. Jessica was buried in the small country churchyard of Sharow, a few miles outside Ripon. Over her grave stands a low pyramid inscribed with the words by which her husband had already described her in that preface:

> In memory of Jessie Piazzi Smyth Daughter of Thomas Duncan the dear Wife of Charles Piazzi Smyth LL. D. Ed. late Astronomer Royal for Scotland, who was his faithful and sympathetic friend and companion through 40 years of varying scientific experiences by land and sea abroad as well as at home at 12000 feet up in the atmosphere on the wind swept Peak of Teneriffe as well as underneath and upon the GREAT PYRAMID OF EGYPT until she fell asleep in the LORD JESUS CHRIST at Clova Ripon on the 24th day of March 1896 aged 80 years.

After his wife's death Piazzi Smyth led an even more secluded life than before though he busied himself on his favourite problems to the end. In his last years alone he was not without support from his and Jessica's numerous relatives and friends. There were the younger members of the Smyth and Baden-Powell families, and the Forbes', Mackenzies and Lumsdens from the network of Jessica's family in Scotland, all of whom were remembered in his will.

The Smyths' tomb at Sharow churchyard. (Photograph courtesy of Rose Coulon.)

Piazzi Smyth survived his wife by four years. He died after a short illness, aged 81, in 1900 and was buried next to her in Sharow. The pyramid over their grave was given a second plaque:

> Beside the Earthly Remains of his Lamented Wife here lies interred the Body of CHARLES PIAZZI SMYTH born 3rd January 1819 died February 21st 1900. Astronomer Royal for Scotland from 1845 to 1888, who earned unperishing renown for his Journeys to distant lands for

scientific objects, and by his eminent Astronomical and other Scientific Writings and Researches. As Bold in enterprise as he was Resolute in demanding a proper measure of public sympathy and support for Astronomy in Scotland, he was not less a living emblem of pious patience under Troubles and Afflictions, and he has sunk to rest laden with well earned Scientific Honours, a Bright Star in the Firmament of Ardent Explorers of the Works of their Creator. Still achieving still pursuing, Learn to Labour and to Wait.
He prayeth best who loveth best All things both great and small
For the dear God who loveth us, He made and loveth all.

As in life, so at the end, Piazzi Smyth kept his affairs neat and in good order. The Royal Society of Edinburgh which he had served well and where he had spent many a happy day, received his valuable sketchbooks, diaries, laboratory workbooks and photographs. To that Society he also bequeathed a legacy for the promotion of research in mountain astronomy, and to the Royal Society of London a sum in repayment of the grants it had awarded him for his spectroscopic experiments at Clova. Much of his scientific and photographic apparatus was unfortunately scattered when the contents of his home at Clova were auctioned after his death, though some of his earlier pieces of optical equipment had been left behind in Edinburgh for the use of the observatory. The beautiful house, Clova, remained in the hands of Jessica's relatives, the Mackenzie family, until 1946.

What of the pyramid? One of Piazzi Smyth's first occupations on his retirement was to bring out the fifth and last edition (1891) of *Our Inheritance in the Great Pyramid* which was still in demand among the pyramid faithful. Compared with the earlier editions this one somehow makes a tired impression. While the early chapters on the value of π and on the sacred cubit remain unchanged, the later chapters are bogged down in abstruse biblical references and quotations from the writings of obscure new disciples. From those pyramidological letters which survive in Edinburgh it is obvious that in the course of time there were disagreements among his followers and conflicts between some of them and Piazzi himself. The latest 'heretic', Flinders Petrie, did not have the warlike character of the old enemies like Simpson or James with whom Piazzi Smyth had waged his early lusty battles. The disagreement with Petrie can have brought no pleasure to Piazzi Smyth, and one wonders whether he secretly suspected that Petrie, whose ability he surely recognised, might be right after all, and that the Earth-commensurable inch was a delusion.

It is not unreasonable to infer a certain change in attitude to the pyramid in Piazzi Smyth's last years. Among the papers left behind in the Royal Observatory are the original letters and manuscripts of

John Taylor, including that of his book *The Great Pyramid; Why Was It Built? & Who Built It?*, the source of all the mischief. Considering the veneration in which the Smyths had held Taylor when they were first converted to his views, one would expect that these papers, entrusted to them on Taylor's death, would have been too precious to part with. Another pointer is the absence of a mention in their last will to any of the groups connected with pyramidology in England or in the United States, to which Piazzi Smyth had occasionally sent donations. Piazzi Smyth's dispositions regarding his worldly wealth tell as well as words where his heart finally lay: it was with his family and friends; with the learned Societies including the Royal Society of London which he had once sacrificed for the sake of the pyramid; and with his first and greatest scientific love, mountain astronomy. In the inscription on the Smyths' tomb, the pyramid, like the Peak of Tenerife, is the scene of a 'scientific experience'. One likes to think of it also as an evocation of the beautiful pyramid-shaped marble tomb of Canova in Venice which had so deeply moved the Smyths on one of their happiest journeys.

The picture of Piazzi Smyth as an apostle of a mystic pyramid cult, which has persisted until the present day, has tended to obscure his real merit as a scientist. There is no doubt that he was an eccentric, with his extraordinary interpretations of the geometry of the pyramid and his fanatical battles against the metric system; but he lived in an age of strong-minded individualists and larger-than-life enthusiasts. The other side of his personality shows a man of many talents, an outstanding observer and gifted experimenter who seized his chance to exercise his skills in spite of a lack of academic training which would have been an unsurmountable obstacle to a less fertile mind. His close friendship with such respected scientists as Peter Guthrie Tait, David Gill and Alexander Herschel illustrates the level of his scientific activities; and the unfailing moral support which he received on behalf of the Edinburgh Observatory from men like William Thomson (Lord Kelvin) and Lord Lindsay proves that he was well regarded by those in the academic world who understood his aims.

In an account of Piazzi Smyth's scientific achievements written after his death, Alexander Herschel (*Nature*, June 1900) dealt in particular with his work in spectroscopy. Using the same flamboyant language which characterized Piazzi's own writings, Herschel called him 'a mighty Ajax in the field of science, a sturdy patient Atlas who through more than half of this fast-waning century robustly upheld on his strong shoulders the growing spires and architraves of science's ever-increasing edifice'. Piazzi Smyth was indeed one of the most successful laboratory spectroscopists of his day, a leading expert in high-resolution work and a specialist in the spectra of low-temperature gases. With instruments which he designed and largely built himself he also made significant contributions to the study of phenomena such as the Zodiacal Light and the Aurora Borealis, and to investigations of the terrestrial atmosphere. His spectroscopic work, to quote Ralph Copeland, his successor in the Edinburgh post, 'will always secure him a high place amongst the scientific workers of the Victorian era'.

Important though these spectroscopic achievements were, his greatest contribution to science has undoubtedly been his pioneering work in mountain astronomy. He was far ahead of his time in insisting on the immense benefit to be gained for astronomy by placing telescopes away from 'the haunts of men' at specially selected locations at high altitude where, to quote his words, 'one night's observation was worth six months in Edinburgh'. He declared as far back as 1852, four years before he carried out his scientific tests on Tenerife, that it was 'strange that men have been content to work under the excessive disadvantage and under the ungrateful difficulties' of dust-laden city sites; yet his vision of a mountain observatory, 'built in the place best calculated according to Newton and Nature', a proposal which to him—and to us today—was so obviously right, came to nothing in his lifetime, or, indeed, what in retrospect seems quite extraordinary, in the lifetimes of several generations of astronomers in the United Kingdom. The blinkered belief that astronomy must be done in the formal setting of an observatory instead of at the place most suitable for the job in hand, persisted all too long.

However, though disappointed at home, Piazzi Smyth lived to see his conviction acted on by the forward-looking new breed of astronomers in the United States. He was delighted when shortly before his retirement he received the news that the Lick Observatory had been opened on Mount Hamilton in California at an altitude of 1280 metres. Edward S Holden, the first Director of Lick, fully recognised Piazzi Smyth's pioneering work to which he devoted the first chapter of his treatise on *Mountain Observatories in America and Europe* (1896). Piazzi Smyth warmly appreciated Holden's references and expressed to him his admiration of the 'superb institution on Mount Hamilton, with the grandest astronomical instruments yet constructed anywhere'.[1] He followed the work of the Lick astronomers with great interest and he no doubt was reminded of his own Tenerife days when he read J M Schaeberle's report on his observations of the solar eclipse of 1893 which he made at the Chilean

Andes from an altitude of 20 000 feet. 'What a privilege', wrote Piazzi Smyth to him, 'to have seen once in your life the mercurial barometer standing at 14 inches only!'[2]

The Americans who not many years before had been novices in the field of astronomy were eager to move ahead as quickly as possible, and near the end of the nineteenth century another distinguished American institution, the Harvard College Observatory, emulated the Lick Observatory by establishing a mountain station of its own. Among the people whom E C Pickering, Director of Harvard Observatory, consulted was Piazzi Smyth who replied to Pickering's enquiries by sending him copies of his Tenerife photographs and who also prepared on behalf of the Royal Society of Edinburgh a 'full statement on Mountain Astronomy' for the benefit of Pickering in 1887. Four years later Harvard established its well-known mountain station at Arequipa in Peru.

Copeland spoke truly indeed when he declared that Piazzi Smyth 'gave Mountain Astronomy an impetus which cannot be overestimated'; and so did Holden when he called Piazzi's station on Tenerife 'the parent of all mountain observatories'.

Notes

Chapter 1

Further details of Piazzi Smyth's years in South Africa and of the history of astronomy in Edinburgh before Piazzi Smyth's time can be found in Warner's and Brück's books (see Bibliography). R L Stevenson's essay on Calton Hill is included in the reprinted *Picturesque Old Edinburgh* with preface by J Desebrock (Edinburgh, 1983). A J Youngson's *The Making of Classical Edinburgh* (Edinburgh, 1966) and the volume on Edinburgh from *The Buildings of Scotland* ed C McWilliam (Penguin, 1984) describe the architecture of the Calton Hill Observatory.

1 J Ashbrook's description in *The Astronomical Scrapbook* (Cambridge, 1984).

2 RS 217, 219 Herschel–Smyth correspondence, Mrs A Smyth to Sir John Herschel.

3 Another sister, Rosetta, married W H (later Sir William) Flower, Director of the Natural History Museum, London. A third sister, Ellen, was the wife of Captain Henry Toynbee, Marine Superintendent of the Meteorological Office. Piazzi's two brothers were both very successful in their chosen professions; both were knighted.

Chapter 2

1 RAS MSS Sheepshanks correspondence.

2 RS 299 Herschel–Smyth correspondence, C P Smyth to Sir John Herschel.

3 In Scotland, a payment made to the superior of a property who may be, for example, the original owner of the land.

4 *Records of the RAS Club* ed H H Turner (Oxford, 1904).

5 D R Hay papers, Edinburgh University Library.

6 Symposium records, Edinburgh University Library; C P Finlayson 'The Symposium Academicum' in G Donaldson (ed) *Four Centuries University Life* (Edinburgh, 1983).

Chapter 3

1 Senatus minutes, Edinburgh University Library.
2 The information in this paragraph comes from Brian Warner's account 'The large southern telescope: Cape or Melbourne?' *Journal of the History of Astronomy* **12** 155 (1981).
3 Henry Clerk papers, Penicuik House, Penicuik, Scotland.
4 B Warner's book (see Bibliography).

Chapter 4

1 'Teneriffe', the old spelling of Tenerife, retained here when quoting the writings of Piazzi Smyth or his contemporaries.
2 A von Humboldt's *Cosmos* translated into English by A Pritchard (London, 1845–8).

Chapter 5

1 Clerk's range-finder (1858) is the earliest one mentioned in L C Martin *Optical Measuring Instruments* (London, 1924).
2 Information kindly supplied by Dr Jean Rösch of the Observatoire du Pic-du-Midi.
3 RAS papers 49.
4 Lee–Smyth letters, Edinburgh University Library.
5 E Neison *The Moon* (London, 1876).
6 Article on gyroscopes by James Blyth in *Encyclopaedia Britannica* (9th ed).

Chapter 6

1 John Taylor was co-publisher of the earlier editions of J Herschel's *Outlines of Astronomy*. He occasionally published for London University but was not related to Richard Taylor, publisher to London University. John Taylor's life is described by Mrs Piazzi Smyth in an Appendix to the 4th edition of *Our Inheritance in the Great Pyramid* (1880).
2 E W Hobson *Squaring the Circle* (Cambridge, 1913).
3 Modern scholars give the date of the building of the Great Pyramid as 2600BC. They no longer consider that the descending entrance passage was intended to point to the pole star, but suggest that a ventilation shaft from the King's Chamber to the northern face of the pyramid pointed to the pole star which at that time was also α Draconis (I E S Edwards *The Pyramids of Egypt* 3rd ed 1985).

The value for the angle of descent of the entrance given by Edwards is 26° 31′ 23″, not very different from the value of 26° 41′ used by Herschel. The tangent of this angle is 0.50, which means that the passage descends with a slope of one part in two, a fact not noted by Herschel who interpreted the angle as representing the altitude of α Draconis at lower culmination in all editions of his *Outlines of Astronomy*.

4 Max Eyth's two-volume book first appeared in 1902; abridged version, Ulm (no date).

Chapter 7

1 The Sothic (from the Egyptian name for Sirius) year was found from the heliacal rising of Sirius. In his lecture notes Piazzi Smyth describes the distinction between this year and the tropical year as obtained from the observation of the equinoxes at the times when two sides of the pyramid pointed to the rising and setting of the Sun.

2 The table shows the geometry of the various models of the pyramid.

Model	Slope (tangent of angle)	Angle	Semiperimeter /height
Perfect π	$4/\pi$	51°51′14.3″	3.14159
Height/side=7/11	14/11	51°50′34″	3.14286
Herodotus formula (area of side=height^2)	$[(1 + 5^{1/2})/2]^{1/2}$	51°49′38″	3.14461
James model (corners rise 9/10)	$9(2^{1/2})/10$	51°50′39″	3.14270

Piazzi Smyth originally claimed that the pyramid measurements could differentiate between the π and the Herodotus models, which meant an accuracy of better than two minutes of arc in the angle and one part in 1000 in the linear measurements. He could not claim an accuracy of half a minute of arc which would be needed to distinguish the James model from the true π model. In fact, his later measurements of an actual casing stone gave limits of 51° 50′ and 51° 53′ for the slope angle, too low an accuracy to distinguish any of the models from each other.

3 O M Mitchell *Orbs of Heaven* (London, 1854).

4 Helen Wright *Sweeper of the Sky, The Life of Maria Mitchell* (Nantucket, 1959).

5 A similar problem, of the structural strength of pyramids, has been discussed in recent times by K Mendelsohn in his book *The Riddle of the Pyramids* (London, 1974).
6 E B Denison *Astronomy without Mathematics* (London, 1865).
7 In Jessie Aitken Wilson *Memoir of George Wilson* (Edinburgh, 1860).
8 P Tompkins's book (see Bibliography).
9 Augustus de Morgan *A Budget of Paradoxes* (London, 1872).

Chapter 8

1 RS R7–66.

Chapter 9

1 The beginning of a new era coincided with the end of another. Brewster and Forbes both died in 1868. Herschel died in 1871.
2 Article on observatories *Encyclopaedia Britannica* (9th ed).
3 A modern scientific observation of the Blue Sun seen in Edinburgh is described by R Wilson in 'The blue Sun of 1950 September' *Monthly Notices of the Royal Astronomical Society* **111** 478 (1951).
4 P Herring 'Lights in the sea' *New Scientist* (22 February 1984). Dr Herring of the Institute of Oceanographic Sciences, Wormley, says that Piazzi Smyth's observations accurately describe dinoflagellates and also that his may be the earliest observation of the spectrum of the luminescence.

Chapter 11

1 Piazzi Smyth admired Le Verrier and gave a commemorative address to the Royal Society of Edinburgh after his death in 1877. He did not hesitate to record Le Verrier's weaknesses while eulogising him in what is described as a 'bizarre *éloge*' in Campbell and Smellie's book (see Bibliography).
2 Sir Robert Anderson KCB *The Lighter Side of my Official Life* (London 1910).
3 *Report of the Edinburgh Royal Observatory Commission* (London, 1876) Royal Greenwich Observatory Archives.

Chapter 12

1 *Memoir of George Wilson*, see reference in Chapter 7.

Chapter 13

1 W H Smyth's work on colours of stars is described in Paul Murdin's article 'Star colours: hit not myth' *Quarterly Journal of the Royal Astronomical Society* **22** 353 (1981).
2 Tait's letter to Thomson mentioning Piazzi Smyth's experiment is quoted in Tait's biography (see Bibliography).
3 Described in J Clerk Maxwell's article on the ether in *Encyclopaedia Britannica* (9th ed).
4 R A Proctor *The Great Pyramid Observatory Tomb and Temple* (London, 1883).
5 Olin J Eggen 'Charles Piazzi Smyth' Astronomical Society of the Pacific, Leaflet 313 (1955) (reprinted in *Monthly Notes of the Astronomical Society of South Africa* **14** 61 (1955).

Chapter 14

1 M Gadsden 'Noctilucent clouds' *Quarterly Journal of the Royal Astronomical Society* **27** 351 (1986).
2 J Halm *Annals of the Royal Observatory Edinburgh* **2** (1906).
3 R A Sampson, *Royal Observatory Edinburgh Annual Reports* (1913).
4 D E Osterbrock *The Astronomy Quarterly* **5** 87 (1985).

Chapter 15

1 Mary Lea Shane Archives of the Lick Observatory.
2 Mary Lea Shane Archives of the Lick Observatory.

Bibliography

Airy W 1896 *Autobiography of George Biddell Airy* (Cambridge)
British Association for the Advancement of Science *Reports*
Brück H A 1983 *The Story of Astronomy in Edinburgh* (Edinburgh)
Campbell N and Smellie R M S 1983 *The Royal Society of Edinburgh, the First Two Hundred Years* (Edinburgh)
Capron J Rand 1879 *Aurorae and Their Spectra* (London)
Clerke, Agnes M 1893 *A Popular History of Astronomy During the Nineteenth Century* 3rd ed (London)
—— 1903 *Problems in Astrophysics* (London)
Copeland, Ralph 1900 Obituaries of Charles Piazzi Smyth in *Monthly Notices of the Royal Astronomical Society* (February) and *Astronomische Nachrichten* **152**
Dreyer J L E and Turner H H 1923 *History of Royal Astronomical Society*, 1820–1920 (London)
Drower M S 1985 *Flinders Petrie, a Life in Archaeology* (London)
Encyclopaedia Britannica 9th ed 1875, 13th ed 1926
Forbes G 1916 *David Gill, Man and Astronomer* (London)
—— 1921 *History of Astronomy* (London)
Grant Sir Alexander 1884 *The Story of the University of Edinburgh During Its First 300 Years* (London)
Herschel A S 1900 Some notes on the late Professor Piazzi Smyth's work in Spectroscopy *Nature* **62** 161
Herschel J F W 1849 *Outlines of Astronomy* (London)
—— 1863 *Two Letters to the Editor of the Athenaeum on a British Modular Standard of Length* (London)
—— 1873 *Popular Lectures on Scientific Subjects* (London)
Kayser H 1900 *Handbuch der Spectroscopie* (Leipzig)
King H C 1979 *The History of the Telescope* (New York: Dover)
Knott C G 1911 *Life and Scientific Work of P.G. Tait* (Cambridge)
McCrea W M 1975 *The Royal Greenwich Observatory* (London: HMSO)
Meadows A J 1970 *Early Solar Physics* (Oxford)
Osterbrock D E 1984 *James E. Keeler: Pioneer American Astrophysicist* (Cambridge)
Petrie W M Flinders 1903 *A History of Egypt* (London)
Petrie, Sir Flinders 1931 *Seventy Years in Archaeology* (London)

Roscoe H E 1873 *Spectrum Analysis* (London)
Scheiner J (trans E B Frost) 1898 *A Treatise on Astronomical Spectroscopy* (Boston and London)
Shairpe J C, Tait P G and Adams-Reilly A 1873 *Life and Letters of James David Forbes* (London)
Taylor J 1859 *The Great Pyramid; Why Was It Built? & Who Built It?* (London)
—— 1863 *The Battle of the Standards* (London)
Tompkins P 1971 *Secrets of the Great Pyramid* (New York)
Secchi A 1875 *Le Soleil* 2nd ed (Paris)
Smyth W H 1844 *A Cycle of Celestial Objects* (including Part 2, *The Bedford Catalogue*) (London)
—— 1851 *Aedes Hartwellianae* (London)
Warner B 1979 *Astronomers at the Royal Observatory Cape of Good Hope* (Cape Town)
—— 1983 *Charles Piazzi Smyth Astronomer–Artist, His Cape Years* (Cape Town)
Wolf R 1890 *Handbuch der Astronomie Ihrer Geschichte und Literatur* (Zürich)

Biographical dictionaries

American *Who's Who* and *Who Was Who*
Biographical Dictionary of American Scientists
Canadian Biographical Dictionary
Dictionary of National Biography
Dictionary of Scientific Biography
Dictionary of American Biography
Monthly Notices of the Royal Astronomical Society Obituary notices

Charles Piazzi Smyth's Publications

Edinburgh Astronomical Observations

Piazzi Smyth's work as Astronomer Royal for Scotland is contained in volumes 7 to 15 of the *Edinburgh Astronomical Observations* which appeared in print between 1868 and 1886. These volumes contain positions of stars observed at the Royal Observatory and the *Edinburgh Star Catalogue*, formal Reports to the Board of Visitors, and meteorological and geophysical records. Piazzi Smyth also used these volumes to publish scientific papers on spectroscopy and meteorology, and the details of his Pyramid metrology.

Scientific Papers

Between 1843 and 1890 Piazzi Smyth published about 100 papers in scientific journals, principally in: *Monthly Notices of the Royal Astronomical Society; Philosophical Transactions of the Royal Society; Proceedings* and *Transactions of the Royal Society of Edinburgh; Transactions of the Royal Scottish Society of Arts*; and *Nature*.

Books

Report on the Teneriffe Experiment of 1856 (London, 1858)
Teneriffe, An Astronomer's Experiment or Specialities of a Residence above the Clouds (London, 1858)
Three Cities in Russia 2 vols (Edinburgh, 1862)
Our Inheritance in the Great Pyramid (London, 1864)
Life and Work at the Great Pyramid 3 vols (Edinburgh, 1867)
On the Antiquity of Intellectual Man (Edinburgh, 1868)
La Grande Pyramide (translated by F Moigno) (Paris, 1875)
Madeira Meteorologic (Edinburgh, 1882)
Madeira Spectroscopic (Edinburgh, 1882)

Articles and pamphlets

Programme of lectures in practical astronomy (Edinburgh, 1853)
New and improved instruments for navigation and astronomy (Edinburgh, 1855)
Present state of the longitude question in navigation (Edinburgh, 1858)
On distance measuring in Russia *Transactions of the Institute of Engineers in Scotland* **3** 96 (1860)
Above the clouds *Good Words* (1862)
Time and its measurement *Good Words* (1862)
Vistas in the Russian Church *Good Words* (1862)
The trophy telescope of the first exhibition *Good Words* (1863)
A poor man's photography at the Great Pyramid in 1865 (London, 1870)
On equal surface projection for maps of the world (Edinburgh, 1870)
Shall we have false or true, earth commensurable measures? (London, 1872)

A complete list of scientific publications by Charles Piazzi Smyth is kept in the archives of the Royal Observatory Edinburgh.

Subject Index

Name Index